# Mathematics Education with Digital Technology

**Other Titles in the Education and Digital Technology Series**

*Drama Education with Digital Technology*, Michael Anderson, John Carroll and David Cameron

*Foreign Language Learning with Digital Technology*, Michael Evans

*Music Education with Digital Technology*, John Finney and Pamela Burnard

**Also available from Bloomsbury**

*Teaching Mathematics Using ICT*, Adrian Oldknow, Ron Taylor and Linda Tetlow

# Mathematics Education with Digital Technology

## Education and Digital Technology

Edited by

Adrian Oldknow and Carol Knights

B L O O M S B U R Y

LONDON • NEW DELHI • NEW YORK • SYDNEY

**Bloomsbury Academic**
An imprint of Bloomsbury Publishing Plc

50 Bedford Square
London
WC1B 3DP
UK

175 Fifth Avenue
New York
NY 10010
USA

**www.bloomsbury.com**

First edition published by Continuum 2011
Paperback edition published 2013



**British Library Cataloguing-in-Publication Data**
A catalogue record for this book is available from the British Library.

ISBN: HB: 978-1-4411-8472-6
PB: 978-0-5672-5028-5
e-PDF: 978-1-4411-8900-4
e-Pub: 978-1-4411-2985-7

**Library of Congress Cataloging-in-Publication Data**
Mathematics education with digital technology /
edited by Adrian Oldknow and Carol Knights.
p. cm. – (Education and digital technology series)
Includes bibliographical references and index.
ISBN 978-1-4411-8472-6 (hardcover) – ISBN 978-1-4411-8900-4 (ebook (pdf))
1. Mathematics–Study and teaching (Middle school)–Technological innovations.
2. Electronic instruments, Digital. I. Oldknow, A. J. (Adrian James),
1946- II. Knights, Carol.

QA11.2.M2779 2011
510.71—dc22 2010033395

Typeset by Newgen Imaging Systems Pvt Ltd, Chennai, India

# Contents

# Notes on Contributors

**Mel Bradford** is a primary school teacher working at a school in Brighton, UK. She has worked in various state schools in the United Kingdom and completed her PGCE at Sussex University in 1991. During her career, Mel has been a co-ordinator for mathematics and assessment. She is currently ICT co-ordinator. As mathematics was her degree subject, she has always been interested in promoting mathematical understanding with primary school children.

**Gail Burrill**, a former secondary teacher and department chair, is currently a Mathematics Specialist in the Division of Science and Mathematics Education at Michigan State University. She served as President of the National Council of Teachers of Mathematics and as Director of the Mathematical Sciences Education Board. She has been involved in using graphing calculators in teaching mathematics since their advent, is an instructor for Teachers Teaching with Technology, and serves as a senior mathematics advisor to Texas Instruments Education Technology. Gail has written and edited many books and articles on teaching and learning statistics and spoken nationally and internationally on issues in teaching and learning mathematics. Her research interests are the use of technology in teaching secondary mathematics, statistics education at the secondary level, and issues related to the professional development of teachers.

**Lupita Carmona** is a native of Mexico. Dr. Carmona has an impressive record of professional leadership and research both in Mexico and the US. Dr. Carmona worked for the Ministry of Education in Mexico in leading national incentives for educational reform through the implementation of new technologies. Later she obtained her doctoral degree in Mathematics Education at Purdue University. Her research focuses on student learning, curriculum development, and evaluation and assessment, with heavy emphasis on collaborative learning and distance education. Dr. Carmona teaches in the UTeach secondary certification programme and the graduate programme in Science and Mathematics Education in the University of Texas at Austin.

**Alison Clark-Wilson** is principal lecturer in Mathematics Education in The Mathematics Centre, University of Chichester, UK. Her research is focused on the development of ICT resources and approaches within secondary mathematics education and, in particular issues concerning teachers' professional development. She has directed a number of research projects in this area. Her publication *Exciting ICT in Mathematics* Network Continuum Education (2005) provides a useful guide for teachers who are developing their use of technology in the mathematics classroom.

**Tina Davidson** is a primary school teacher currently working at a school in Brighton, UK. She has worked in various state nursery, infant and primary schools in England and spent a year teaching in Australia. She qualified as a teacher in 1986 with a Bachelor of Education degree with Honours. During her career, Tina, has co-ordinated ICT, Literacy, Assessment, Early Years, Self-evaluation and currently mathematics. Tina enrolled on a part-time MA in Mathematics Education in 2008 at the University of Chichester where she developed an interest in developing children's mathematical ability through the use of ICT.

**Sarah M. Davis** is a faculty researcher at the Singapore Learning Sciences Lab at (NIE), an institute of Nanyang Technological University. Dr. Davis received her undergraduate degree in communications from Concordia University and her masters and doctorate in mathematics education from the University of Texas at Austin. Between her undergraduate work and beginning her doctorate, she was a classroom teacher for 6 years, having taught regular, special education and gifted mathematics at both the middle school and high school levels. In addition to teaching mathematics, Dr Davis also taught 1 year of regular education third grade. More recently she's had years of experience introducing and supporting near-term innovation in schools serving at-risk populations, substantial background working with teachers in both pre-service and in-service settings across the US, extended research participation in the Participatory Simulations Project funded by the National Science Foundation (NSF), an effort funded by the NSF to research and develop – with Texas Instruments – a new, wireless, networked classroom technology aimed at transforming the classroom as a dynamic learning environment. During a leave of absence from the University of Texas, Dr Davis worked for Texas Instruments, Inc. on the creation and release of the TI Navigator ™ classroom network. She co-ordinated the pilot site community working with 60 teachers at 30 schools in 20 states, making sure that the teachers using the pre-release version of the system were supported and successful in using the technology. During her last year at TI she worked as a Project Manager, organizing multiple development teams to facilitate the initial release of the TI Navigator product. Currently, extending her dissertation work, she is developing a research agenda focused on function-based algebra, classroom networks, generative activities and their design, teacher professional development, group dynamics and the use and affordances of anonymity in the classroom.

**Rosemary Deaney** is a senior teaching associate in the Faculty of Education, University of Cambridge. With a background in teaching across a range of educational sectors, she has a research interest in the use of digital technologies in subject teaching and learning. She has worked as Research Associate on several collaborative projects in this area, including supporting teachers developing their practice with technology tools.

**Dawn Denyer** has 13 years experience of teaching in the south of England. She started her teaching career in a large 'traditional' comprehensive, and has since taught or led in Specialist Technology Colleges. She is currently Mathematics Subject Leader in a designated Arts College. Dawn enjoys using technology in her classroom to engage and motivate her students, and encourages other teachers to do the same.

**Bryan Dye** is a former mathematics teacher of 28 years experience in comprehensive schools, including a period as advisory teacher for Norfolk. He has authored published teaching resources for spreadsheets and Omnigraph and for about 7 years contributed the WebWatch column to ATM's termly magazine. Since the late 1990s he has developed interactive websites under the MathsNet name and is currently working on advanced mathematics www.mathsnetalevel.com/, a site catering for students and teachers of A-level, International Baccalaureate and other similar courses.

**Jim Fensom** is a mathematics teacher at an international school: the United World College of South East Asia, Singapore. He graduated from Southampton University with a BSc degree in Mathematics and subsequently a part time MEd degree from Sheffield University. With over 30 years of experience of teaching mathematics at all levels in the International Baccalaureate, Jim has been consulted by examination boards and calculator manufacturers. He specializes in the effective use of technology in the teaching of mathematics such as graphing calculators, data loggers and electronic whiteboards.

**Ian Galloway** is Deputy Director at the Science Learning Centre South East based in the University of Southampton. He is also Education Director for BLOODHOUND SSC and sees this type of project as a great hook on which to hang children's learning. Ian is very keen to promote the place of mathematics within science education and believes that it is not possible to offer a coherent science education to children without integrating mathematics. He worked within the state and private sectors for over 30 years before moving into teacher training and professional development.

**Ghislaine Gueudet** is full professor at IUFM Bretagne (teacher training institute). She is vice-director of the CREAD (Center for Research on Education, Learning and Didactics), and member of its ICT research axis. Her interest for ICT started with the didactical study of the use of e-exercises bases, from grade 3 to mathematics master degree. It led her then, in a joint work with Luc Trouche, to consider all the resources intervening in the teacher's activity. They introduced an approach for the study of mathematics teachers' documentation work. They still work on the development of this approach, and simultaneously implement it in different research projects.

**Peter Hamilton** is the Head of Education Development with Intel Performance Learning Solutions. Peter has over 25 years experience in the IT Industry and has been with Intel since 1991 where he has worked in a in a variety of manufacturing, quality management, and technology development roles in both Ireland and the US. In 1998 Peter was a founder member of Intel Performance Learning Solutions. He has led the education research, design and development activities which have delivered the award winning **skoool™** Learning and Teaching technologies. This is now a worldwide programme which provides valuable learning and teaching resources, technologies and strategic approaches in 7 languages and is now operating in more than 30 countries. Peter has consulted on emerging education technologies in North America, Europe, the Middle East, Africa and Asia and has published papers on mobile and technology-based learning. Peter has a Masters Degree in Engineering from University College Dublin. In 2004 Peter was joint recipient of the Digital Media Person of the Year Award at Ireland's National Digital Media Awards for his work with the 'skoool' Learning and Teaching Technologies.

**Michael Hartnell** is Head of Department in an 11 to 16 school in Southern England. He has been involved with a number of local and national developmental projects and has appeared on Teachers' TV showcasing lessons which utilize ICT in contextualized problem solving activities.

**Steve Hearn** is a physics teacher and housemaster at Charterhouse school. He has been involved in developing novel physics investigations using computer modelling, video and data logging as well as rulers and blu-tac! He is also an Institute of Physics network co-ordinator in Surrey, an SEP teacher fellow and he runs the summer residential Subject knowledge booster courses in Chemistry and Physics at Charterhouse.

**Sara Hennessy** is a lecturer in Teacher Development and Pedagogical Innovation at the University of Cambridge. Her research over the last couple of decades into subject teaching and learning using a range of classroom technologies seeks to understand and develop pedagogy, especially related to mathematics and science education. She has recently explored the potential of the interactive whiteboard to support classroom learning through dialogue. Her work also focuses on research partnerships and on practitioner-led professional development in UK and African schools.

**Pip Huyton** is an Independent Mathematics Advisor. Since 1996 her work in Mathematics for Local Authorities, Ark Academies, Becta, the Mathematical Association and the Department for Children Schools and Families has included Professional Development provision for teachers and project management. She is recognized as meeting the National Centre for the Excellence in the Teaching of Mathematics (NCETM) Quality Standard for CPD providers and is an NCETM Associate. Pip has a particular interest and expertise in the effective use of ICT in the teaching and learning of Mathematics, and wide experience in developing collaborative networks of teachers.

**Keith Jones** is Associate Professor of Mathematics Education at the School of Education, University of Southampton, UK, where he is head of the university's Mathematics and Science Education Research Centre. His research expertise encompasses the acquisition and use of mathematical knowledge in different cultural settings, the development of mathematical reasoning in students, and the integration of technology in the teaching and learning of mathematics. He has published widely, including co-editing complete volumes on technology in mathematics education for the *International Journal of Technology in Mathematics Education* and the international journal *Educational Studies in Mathematics.* He has been a member of the thematic group on *Tools and Technologies in Mathematical Didactics* of the *European Society for Research in Mathematics Education* (ERME) since its inception, and, from 2000–2003, he led the group. For up-to-date information, see: www.crme.soton.ac.uk.

**Andy Kemp** is a secondary school mathematics teacher and is currently Head of Mathematics at Taunton School, Somerset, UK. He studied his

Mathematics degree at Warwick University later returning there to complete his PGCE in 2005 and then a part-time MSc in Mathematics Education in 2008. His academic interests focus on the appropriate use of technology in the mathematics classroom, with his MSc research being related to computer algebra usage within secondary schools. He is particularly interested in the potential impact of the internet on mathematics education.

**Carol Knights** is a Principal Lecturer in mathematics education at the University of Chichester, teaching on undergraduate and Masters level courses. She gained experience as a teacher working for 14 years in Hampshire as Head of Department and Advanced Skills Teacher, teaching across the 11–18 age range and working with a range of schools to improve attainment and engagement in mathematics. She has wide expertise in the use of ICT in the classroom and has authored resources for both the Bowland Key Stage 3 mathematics initiative and the GE STEM *Achievement in Mathematics* London Pilot. She currently leads the Chichester team in co-ordinating the work of the NCETM in the South East Region.

**Colette Laborde** is currently Professor Emeritus at the University Joseph Fourier, Grenoble, France. Her research work in mathematics education deals with the integration of the computer in the teaching and learning of geometry and she is involved in the project Cabri-géomètre, dynamic geometry software programs for plane geometry and 3D geometry, distributed across the world. She was the co-chair of the Topic Study Group on 'New technologies in the Learning and Teaching of Mathematics' at the 11th International Congress in Mathematics Education (ICME11) in July 2008. She edited several books in mathematics education and she is a member of several Editorial Boards of International Journals in Mathematics Education

**Ed Laughbaum** is an emeritus professor of mathematics, and is currently the director of the Early Mathematics Placement Testing Program at The Ohio State University, US. He is presently interested in the implications of basic brain processes on understanding and memory/recall as related to the teaching of algebra with handheld technology. His textbook *Foundations for College Mathematics* represents a first iteration of an implementation. Ed has authored over 50 publications in professional journals or by commercial publishers. He has given over 250 presentations at state, national, and international conferences.

**John Mason** is professor emeritus at the Open University and Senior Research Fellow at the Department of Education in the University of Oxford. His interests are centred on fostering and sustaining mathematical thinking, and supporting those who wish to do likewise. He adopts an explicitly experiential way of working, promoting engagement in mathematics together with reflection on the actions and processes, and mental preparation for acting in an informed manner in the future. He has authored and co-authored numerous books and pamphlets for teachers, including *Thinking Mathematically*, *Mathematics as a Constructive Activity*, *Designing and Using Mathematical Tasks*, and *Fundamental Constructs in Mathematics Education*.

**Adrian Oldknow** is Emeritus Professor of Mathematics and Computing Education at the University of Chichester. He was Open Scholar in mathematics at Hertford College, Oxford University, where he obtained his MA in mathematics. After teaching mathematics in grammar, comprehensive and independent schools he moved to a lectureship in mathematics and computing at Reading College of Technology, and took a part-time MTech degree in Computer Science at Brunel University. He was Head of Mathematics at the Chichester Institute of Higher Education, and awarded a Personal Chair at University College, Chichester. He has held fellowships at King's College London, the University of London Institute of Education and the University Putra Malyasia. He is currently research fellow at the Amstel Institute, University of Amsterdam. A past editor the IMA journal *Teaching Mathematics and Its Applications*, Adrian has written extensively on the impact of ICT in teaching and learning mathematics. He now works as a free-lance consultant, and has a particular interest in STEM.

**Chris Olley** is the director of the secondary mathematics PGCE course at King's College, London. He has extensive teaching experience in inner city comprehensive schools where he embedded the use of graphing calculators and dynamic mathematics software at all levels. He is currently engaged in research on the variation in take up of different types of interactive mathematics ICT tool amongst PGCE students in the course of their training. His PhD, in progress, is about engagement with dialogic relations in secondary mathematics classrooms.

**Don Passey** is a senior research fellow in the Department of Educational Research at Lancaster University, UK. His research, for government departments and agencies, non-commercial and commercial groups, has focused on the identification of teaching and learning outcomes arising from uses of leading edge technologies, how home and out-of-school practices enhance and support learning at an individual pupil level, how technologies support specific groups of young people (including those at risk), and how evaluation and research can be undertaken to support policy and practice.

**Matt Pauling** is a National Development Manager for the Youth Sport Trust, working with specialist sports colleges to use PE and sport to raise whole school standards. He has led on the core subjects' programme for the YST supporting using sporting contexts, values, skills and pedagogies to improve motivation, achievement and attainment in English and maths. As a former PE teacher and keen sportsman he is passionate about the power sport can have on the engagement and aspirations of young people and the added value that relevant and accessible new technologies can bring to learning.

**Vanessa Pittard** is Director of e-Strategy at Becta, leading Becta's work in the areas of strategy, research, and technology innovation and futures. Originally appointed as Director of Evidence and Evaluation, Vanessa has led Becta's research since 2004. Prior to working at Becta, Vanessa led the ICT Research and Evaluation team at DfES, developing and managing a programme of research to inform the development of technology policy and strategy. Before 2002, Vanessa had a long career in the University sector, leading the Department of Communication Studies at Sheffield Hallam University before moving into government research.

**Sue Pope** is programme manager for mathematics at QCDA. She believes that all learners should have the opportunity to experience mathematics as the exciting and creative subject that it is. She moved to QCDA after 10 years in higher education at the University of Surrey Roehampton and, most recently, St. Martin's College, Lancaster. During that time she worked with beginning primary and secondary teachers on undergraduate and postgraduate courses, and supported experienced teachers working towards higher degrees. For a short time she worked as a local authority adviser after 10 years teaching in a number of schools, including 5 years as head of mathematics in an 11–18 mixed comprehensive.

**Russell Prue** is a well-known ICT Evangelist and popular conference speaker at education events in the UK and Europe. He is passionate about the use of ICT in all subjects and is responsible for the invention of the cre8txt keyboard a T9 USB keyboard for reluctant writers. Russell spends most of his time motivating and inspiring learning and teaching colleagues to use technology in a learner centric fashion. He is the author of the presenters' handbook *The Science of Evangelism* (2005) and his most recent project is the creation of an innovative live school radio broadcasting system.

**Walter Stroup** serves as an Associate Professor of Science, Technology, Engineer and Mathematics Education at The University of Texas at Austin. Much of his research and on-going school-based implementation efforts focuses on a programme of interdisciplinary design and development fusing the early learning of powerful ideas in mathematics, science and systems theory with advanced technology design. His work with highly interactive networks in school classrooms is supported by a theory of generative design that attends to the senses in which participation in mathematics and science is both socially structured and socially structuring.

**Ruth Tanner** is a secondary mathematics consultant with Shropshire Council. She is an enthusiastic user of many different forms of ICT in the classroom and is particularly interested in the creative use of ICT to explore mathematical concepts. As a member of both the MA and ATM she helped to found the East Midlands joint branch and is currently an active member of the Marches joint ATM / MA branch and of the ATM ICT task group.

**Ron Taylor** is a visiting fellow at Southampton University and a mathematics adviser. He has recently retired from his post as Hampshire inspector/adviser for mathematics. During his 19 years in Hampshire he has been responsible for a number of curriculum initiatives of national significance, and is, by invitation, an adviser on two curriculum working parties under the auspices of the Mathematical Association (MA) and British Educational Communications and Technology Agency (Becta). This has enabled Ron to secure funds for a number of projects based in Hampshire schools. He was a member of the Qualifications and Curriculum Authority's post-16 mathematics advisory group, and was involved in the formal consultation of the review of the mathematics National Curriculum, including, classroom-based research on developing pupils' algebraic and geometrical reasoning. He has been called upon to make representations on proposals for the future curriculum and assessment for 14–19 year olds. Publications include joint authorship of *Engaging Mathematics,* written for the Technology Colleges Trust (now Specialist Schools Trust), and key contributions to *ICT and Mathematics: A Guide to Learning and Teaching Mathematics 11–19* produced for the Teacher Training Agency by the Mathematical Association. He was, by invitation from HMI, part of the team evaluating the impact of the government initiative on the use of ICT in schools. Ron has been particularly keen to develop teacher networks as a means of disseminating effective practice.

**Linda Tetlow** has many years experience in mathematics teaching both in secondary and further education in the south of England. This includes being Head of Mathematics in an 11–18 school, A-level Mathematics coordinator in a Further Education college and Mathematics coordinator for an out-of-school learning-on-line project. She has written activities for GCSE and A-level to encourage the use of ICT in mathematics and has worked as a consultant on projects for the Mathematical Association, the QCDA and the NCETM. She is now an independent education consultant. She has a BSc degree in mathematics from Hull and an MA in Mathematics Education from Chichester.

**Luc Trouche** is full professor and head of the ICT and Education department in the French National Institute for Pedagogical Research (INRP). His field of research concerns the interactions between teachers' development and resources development. His interest in resources started with the didactical study of the conditions of ICT integration in mathematics classes, the resources design and teachers' training required by this integration. It led him then, in a joint work with Ghislaine Gueudet, to consider all the resources intervening in the teacher's activity. They introduced an approach for the study of mathematics teachers' documentation work. They still work on the development of this approach, and simultaneously implement it in different research projects.

**Pat Woolner** is a research associate in the Research Centre for Learning and Teaching at Newcastle University, UK. Formerly a secondary school mathematics teacher, she has carried out research into forms of representation in learning and teaching mathmatics. Other research in which she has been involved has included investigations of the school learning environment and evaluations of a range of school-based learning innovations. Her current role also involves supporting practising teachers from primary, secondary and further education in action research projects.

**David Wright** is programme leader for secondary mathematics initial teacher education at Newcastle University. He taught in secondary schools and further education for 15 years before joining the British Educational Communications and Technology Agency (Becta) as subject officer for mathematics. He was also co-editor of Micromath, a journal of the Association of Teachers of Mathematics. He is particularly interested in the transformative potential of ICT for mathematics education when integrated into the mathematics classroom through the use of connected portable technology to form an interactive social space for doing mathematics.

# Series Editor's Foreword

This is a book about innovation in teaching mathematics using digital technologies, and takes an innovative approach in presenting a range of perspectives, taken from national and international research, to explore key questions about teaching and learning in mathematics.

In this exciting collection of voices, Oldknow and Knights have scaffolded a conversation between well known and respected international writers and researchers in the field, in ways which allow readers opportunities not only to engage with the chapters in conventional ways, but also invite the reader through skilful questions provided in the introduction to each section, to be a third party in discussions between authors, as they themselves offer differing views on such exciting areas as mathematics and neuroscience, intercultural constructions of mathematics teaching, and the possibilities of home-school enrichment through digital technologies.

The scope of this volume is comprehensive. The first section, *Where are we now?* offers a fascinating set of accounts which present an international overview of the current and future possibilities of mathematics and technology: learning from other countries is a hugely valuable way of developing our own self awareness, and the developing areas presented there are a valuable contribution to pointing up how technology has impacted on teaching and learning internationally. The next section, *What does research tell us?* uses research to deepen our understanding of where technology enhanced mathematics teaching impacts on student learning; the well known scholars reporting their research here allow us to see some of the cutting edge thinking which underpins development; this section is cleverly balanced with three further sections focusing on *practical use of ICT in* classrooms, including the intriguing Pizza Problem, and of examples of teaching and learning in mathematics through a series of *case studies* which tell us about real classroom uses, and with real classroom thinking thoughtfully explored. The possibility of technology 'minding the gap' between *school and home* is explored through the uses of personal portable technology whilst the *pedagogy of digital technology in mathematics* is addressed through three chapters which are tied together through John Mason's notion of engagement through manipulatives.

Mathematics is not simply presented as a single, timetabled subject, though. Its wider role in *developing other subject areas*, including sports and leisure, physics and the wider role of STEM is addressed through a selection of accounts which demonstrate the centrality of mathematics to knowledge itself. The holistic approach to thinking about mathematics and digital technology these sections offer is supported by a final section which considers the *role of professional development* in bringing about the types of practices explored in this book through grounded guidance on teacher education.

Oldknow and Knights' vision for this volume in inviting critical engagement with, and development of, new practice within the field of mathematics and digital technologies through the presentation of a series of linked, focused and thought provoking chapters is itself stimulating. I am delighted to be able to include this book in the series on teaching with digital technologies, and my thanks go to both Adrian Oldknow and Carol Knights for this excellent volume.

Sue Brindley

# Acknowledgements

The editors would like to express their gratitude to all of the contributing authors, without whom this book would not have been possible.

Thanks also go to the following individuals, companies and organizations who have granted permission to use screen shots from software, photographs and other images: Doug Brown, CabriLog, Jen-Chung Chuan, Data Studio, Doyle Davis, Vitor Duarte Teodoro, Bryan Dye, MyMaths, Chris Olley, Matt Pauling, TDA, T-media, Texas Instruments, Kingsley Webb, Vernier, Video Point and Phillip Yorke.

Thank you to all the staff at Bloomsbury who worked with us on this book.

Finally, thanks to colleagues Afzal Ahmed, Alison Clark-Wilson and Anne Harvey at the University of Chichester, and also Linda Tetlow and Ron Taylor for their various forms of support and advice.

# Introduction

Digital technologies – especially portable ones – are universal and affect all walks of life. They allow people to communicate anywhere, anytime; to take photographs and movies; to record and play music and films; to play games; to gather information via the internet; to find out where they are and how to get to where they want to be; to compose, send and print documents; to keep their finances in order, and so on. Schools, as with all organizations, rely heavily on the technology for their administrative functions – and most teachers have received training in how to use common elements (internet, e-mail, word-processing, spreadsheets, presentation and display software, learning platforms etc.) to improve their efficiency as teachers. The big challenge to education over recent years has been how to integrate digital technologies into the teaching and learning of subjects to the benefit of all. This is clearly not a challenge which can be met by a "one-size fits all" solution, and so this series of books from Bloomsbury comes at a very important time when many learning communities are reviewing their approaches to education.

In mathematics education we have seen a radical change in the range of uses of digital technologies to support teachers in the classroom, and to enable students to access school work from home. In the United Kingdom most secondary school (students aged 11–19) mathematics teachers have ready access to computers (usually laptops), projectors and interactive white-boards. All schools now have their own websites and learning platforms from which teachers, students and parents can access information about the timetable, curriculum, examinations, careers etc. as well as exercises, homework and so on. There is also usually support for messages between students and teachers.

Very recently we have seen a number of educational 'ultra-mobile personal computer' devices being developed specifically for student use – robust, lightweight, wireless, long-lasting batteries, powerful and cheap. These include MIT's OneLaptopPerChild (OLPC), Intel's Classmate, and now an Indian educational tablet PC for $35. Very soon we can expect to see something resembling the current Apple i-Pad in the hands of learners all over the world. So we need to prepare for a new era in which students do not just have access to an electronic calculator, but the potential to run applications which can support all forms of mathematical and scientific computation, as well as to program their own.

So, if we are to make best use of the potential of such digital technologies to radically improve teaching and learning mathematics, we need to take stock of what we know now, what we have seen to work, what we would like to see tried as well as how to introduce, support and sustain the innovations required. There is also an added political imperative in many nations now for schools to place greater emphasis on the STEM subjects (Science, Technology, Engineering and Mathematics) to ensure more school-leavers head either for Higher Education courses in these disciplines or directly for employment in which these are key aspects. So, like it or not, we are likely, as mathematics educators, to be put under increasing pressure to provide more relevant skills for the national economy.

This book has given us the opportunity to ask experts from around the world to share their experiences and views. These are organized under the following headings.

'Part 1: Where Are We Now?' reviews the current state of play in education with digital technologies in United States, France and the United Kingdom, together with a review of the underlying neuroscience.

'Part 2: What Does Research Tell Us?' draws messages and lessons from some major international projects.

'Part 3: Key Pedagogical Issues in Embedding ICT in Teaching and Learning Mathematics' invites leading educational experts to explore in depth some facets of the impact of digital technologies on learning mathematics.

'Part 4: Description of a Range of Important ICT Tools' identifies important features for mathematics education in the developing technologies.

'Part 5: Practical Ideas of ICT to Enhance Teaching and Learning' highlights examples of interesting recent educational developments with the technology.

'Part 6: ICT Supporting Cross-curricular Work with Mathematics' uses examples from physics, engineering, technology and sports to show how digital tools make data capture and modelling more accessible and relevant, and hence support a joined-up curriculum.

'Part 7: Case Studies of Teachers Engaging with ICT' asks classroom teachers to share their experience, reflections and observations of classroom innovation with digital technologies.

'Part 8: Implications for Professional Development' addresses the important issues arising from the need to introduce, support and sustain educational innovations to a wider teaching workforce.

This is an exciting and challenging period in education at all levels. Despite current economic difficulties, there are many factors which can support the harnessing of the potential of digital technologies to make radical improvements in teaching and learning mathematics. We hope that the sound experience, advice and comments of our authors will make a positive contribution to realizing that potential.

Adrian Oldknow and Carol Knights

# Part One

# Where Are We Now?

In this section we have contributions from France, the United Kingdom and the United States reviewing the extent to which digital technologies have become embedded in their respective educational systems, and what trends are now emerging. We have not attempted to get a systematic international survey as it is impossible to discount factors such as – the different attitudes shown by different cultures towards education and mathematics, how centralized the political system is, how schools are managed, what objectives the educational process is seen to serve, and so on. What does seem apparent is that the digital divide between teachers who have access to Information and Communications Technology (ICT) to support their role as a teacher and those who don't has undergone a rapid change. This has occurred through access to devices such as broadband, the internet, portable computers, digital projectors, interactive whiteboards and virtual learning environments. We can see that teacher groups are developing curriculum resources which are being put into widespread use through digital distribution – even if the result is just a printed worksheet! What is emerging now, with the introduction of ever more portable, powerful and affordable student devices, is the challenge to integrate these into mathematical learning, and assessment. With regard to widespread students' own use of technology as a research tool, an analysis tool and/or a presentation tool, it looks, in the words of Lewis Carroll, like a case of 'jam to-morrow'. But we can see some very interesting approaches developing, and we can hope that, unlike the White Queen's interpretation, 'to-morrow' may not be long in coming. Which is why it is also helpful to include a contribution reviewing what we know about how students learn.

Chapter 1

# The Neuroscience of Connections, Generalizations, Visualizations and Meaning

Edward D. Laughbaum
*Department of Mathematics, The Ohio State University, US*

## Introduction

Did you ever wonder whether teachers consider basic brain function of students when designing lessons or lectures? That is, do we ever think about teaching to be in concert with how the brain functions? Have we considered capitalizing on basic brain function that will improve understanding and long-term memory with recall? Are we aware that the brain requires neural connections to process understanding, long-term memory, and recall? Do we know whether the brain commonly generalizes through reasoning or through pattern recognition? Do we know why it is important for students to generalize a pattern on their own? Have we thought about whether using visualizations to *confirm* mathematical processes and concepts holds the same understanding/memory value as does using visualizations to *teach* processes and concepts? If we knew the answers to these questions, would we change the way we teach? Would textbooks change to facilitate such teaching? Would standards documents focus on teaching instead of providing topic lists?

There are a considerable number of basic brain operating functions that can be applied to the field of mathematics education, but in this chapter, the author will only reference research in brain function related to connections, pattern recognition, visualizations and meaning. Some may question the validity of a proposal to change education to be in concert with basic brain function since the brain is so complex. It is complex, but the proposal to change teaching is based on common neural function (no matter how complex on a cellular and molecular level), and the implementation is somewhat simple. Breakthrough ideas often emerge by applying ideas from one field to another.

## Connections in Mathematics = Associations in the Brain

Neural associations are the connections among neural networks that the brain creates *automatically* and *instantaneously* when it learns something new. Donald Hebb discovered the creation of associations over 50 years ago. We commonly describe his discovery as 'Neurons that fire together, wire together'. For example, suppose you want to create neural associations among the numeric, graphic and symbolic representations of a function. It is extremely simple to do. Graph a function on a graphing calculator (or computer) and use trace (with expression turned on) to trace on the graph, or use a graphic/numeric split screen. Both of these options present the brain with the simultaneous representations of a function causing the neural networks for the three representations to be associated (connected). But why is this important?

Current research shows that ' . . . the lower left part of the frontal lobe works especially hard when people elaborate on incoming information by associating it with what they already know' (Schacter, 2001, p. 27). But there are issues when we facilitate associations in maths education. 'This echo [neurons continuing to fire after the stimulus has stopped] of activity allows the brain to make creative associations as seemingly unrelated sensations and ideas overlap' (Lehrer, 2009, p. 130). Do we really want students connecting addition of polynomials, for example, to concepts that are unique to each student? Doesn't it make more sense for the teacher to facilitate the creation of appropriate associations that can be used later in the teaching/learning process? It is possible. 'Being able to hold more information in the prefrontal cortex, and being able to hold on to the information longer, means that the brain cells are better able to form useful associations' (Lehrer, 2009, p. 131). Outside of education, it is common to lead an audience to connections of choice. For example, 'Advertisers don't wait for you to develop your own associations. They go ahead and program you with theirs through television [like a cool-looking person smoking, or females showing interest in guys in cars]' (Brodie, 1996, p. 25). Of course, what politician has not used the word 'trust' on the same TV screen with their name? Recall that it is rather simple to create associations. Simultaneously present the brain with the concepts/procedures you want connected. But again, why are connections important?

Teachers must create connections to improve the memory of the mathematics taught. 'Memory recall almost always follows a pathway of associations. One [neural] pattern evokes the next pattern, which evokes the next pattern, and so on' (Hawkins, 2004, p. 71). In teaching factoring of polynomials, one would connect the new maths being taught to the previously taught concept of zeros of a function. Using hand-held or computer technology, it is relatively simple to find zeros of polynomial functions expressed as rational numbers. By connecting the two processes, when students are asked to factor a quadratic polynomial at a later time, they are likely to think of zeros first (because of the visual methods used in teaching), followed by the factoring process.

> The most important property [of auto-associative memory] is that you don't have to have the entire pattern you want to retrieve in order to retrieve it. . . . The auto-associative memory can retrieve the correct pattern, . . . even though you start with a messy version of it. (Hawkins, 2004, p. 30)

So we have good odds that *connected* concepts will be recallable.

Teachers must also create connections to enhance the understanding of the concept or procedure being taught. That is, 'We understand something new by relating it to something we've known or experienced in the past' (Restak, 2006, p. 164). The word understanding seems to hold value in the minds of many educators. For example from Keith Devlin,

> How many children leave school with good grades in mathematics but no understanding of what they are doing? If only they understood what was going on, they would never forget how to do it. Without such understanding, however, few can remember such a complicated procedure for long once the final exam has ended. . . . What sets them [those who 'get it'] apart from the many people who never seem to 'get it' is not that they have memorized the rules better. Rather, they understand those rules (2000, pp. 67–68)

We will find that visualizations and pattern recognition also contribute to the understanding of mathematical procedures and concepts. They are discussed below.

Mathematical connections typically come in two forms. The first and most important connection is to previously taught mathematics, but also, 'New information becomes more memorable if we "tag" it with an emotion [like a familiar real-world context]' (Restak, 2006, p. 164). So we also need to connect new maths concepts to contexts that are familiar (evoke an emotional response) to students. For example, when *teaching* (not applying) the concept of the behavior of zero(s) of a function by modelling the amount of fluid remaining in an I.V. drip bag, we 'tag' it with the real-world meaning of the zero – the bag is empty. That is, the nurse must take action at the zero. If the nurse does not replace the bag, the patient may die. If the patient dies, the zero becomes important to the prescribing doctor and several lawyers, and so on. The result of tagging a mathematical concept or process with an emotional connection is improved memory. It turns out that the more connections to a mathematical concept/procedure, the more likely the correct recall. That is:

> In general, how well new information is stored in long-term memory depends very much on depth of processing, . . . A semantic level of processing, which is directed at the meaning aspects of events, produces substantially better memory for events than a structural or surface level of processing. (Thompson and Madigan, 2005, p. 33)

## Pattern Building to Pattern Generalizing

Using pattern building as a tool to help students generalize a pattern, like for example the first law of exponents, has a stained history. The pervasive view is that mathematics is understood through 'reasoning', and this is the standard to which mathematicians typically hold. This may be a noble thought, but it turns out that reasoning is NOT the brain's dominate mode of operation. Gerald Edelman is a Nobel Laureate in medicine and makes an interesting point, 'human brains operate fundamentally in terms of pattern recognition rather than logic [reasoning]. It [pattern recognition] is enormously powerful, but because of the need for range; it carries with it a loss of specificity' (Edelman, 2006, pp. 83, 103). Of course, this loss of specificity is what concerns educators. In mathematics, we may want students to generalize the exact concept/procedure of our choice, and not other options that are open to the student's brain. Yet given the evidence that the primary mode of operation of the brain is pattern generalizing; shouldn't we capitalize on this? Might it improve understanding and memory? A good option for implementing pattern building is to use guided discovery activities because ' . . . the use of controlled scientific observation enormously enhances the specificity and generality of these interactions' (Edelman, 2006, p. 104). Based on the author's experience, successful guided discovery activities are short and lead directly to the desired mathematical generalization. The average brain will generalize on the third iteration. As you might expect, some students will generalize after the first or second – especially after using the process in class for a while, so one needs to think through the guided discovery activity questions that lead students to generalize. However, 'After selection occurs . . . refinements can take place with increasing specificity. This is the case in those situations where logic or mathematics can be applied' (Edelman, 2006, p. 83). Thus, it seems that guided discovery is a good choice.

There is more to the idea of tapping into common brain function through pattern building. We may not know at what point, if any, that the brain creates a long-term memory of a mathematical procedure through practice. But, in pattern building, 'If the patterns are related in such a way that the [brain] region can learn to predict what pattern will occur next, the cortical region forms a **persistent representation, or memory**, for the sequence' (Hawkins, 2004, 128, emphasis introduced). It is crucial that teachers not 'tell' students the desired generalization, but teachers must structure the pattern-building activity that gently leads students to generalize after a reasonable pattern-building activity has been completed. An excellent tool for knowing what each student has generalized is the TI Navigator. With the proper use of Navigator, every student must generalize, and not just the student holding up their hand.

As soon as the student makes the generalized pattern, we know the memory has been created. In addition, the recognition of the pattern by students activates the neural reward system, ' . . . which is consciously experienced as a **feeling of knowing**' (Burton, 2008, p.135, emphasis introduced). This feeling of knowing promotes the flow of the neuro-transmitter dopamine. 'This system is found in the basal ganglia and the brain stem. The release of dopamine acts as a reward system, facilitating learning' (Edelman, 2006, p. 31). 'Our brain with its capabilities of pattern recognition, closure, and filling in, goes, as Jerome Bruner pointed out, beyond the information given' (Edelman, 2006, p. 154). The bottom line is that we should facilitate the brain's ability to generalize patterns, not just for the mathematics, but for all of life where generalizing correctly is of utmost importance.

## Visualizations

The idea that our vision system navigates through our mathematical thinking may seem unusual, but the fact is that 'Neuroplasticity, . . . can reshape the brain so that a sensory region performs a sophisticated cognitive function' (Bagley, 2008, p. 99).

> Mathematical reasoning both takes from and gives to the other parts of the mind. Thanks to graphs, we primates grasp mathematics with our eyes and our mind's eye [occipital lobes]. Functions are shapes (linear, flat, steep, crossing, smooth), and operating is doodling in mental imagery (rotating, extrapolating, filling, tracing). In return, mathematical thinking offers new ways to understand the world. So, vision was co-opted for mathematical thinking, which helps us see [understand] the world. (Pinker, 1997, pp. 359–60)

Processing mathematics through the visual system suggests that we should integrate visualizations in our pedagogy. But there is more . . .

Instead of accepting the reasoning above as appropriate, might there be another rationale for using visualizations? The author suggests at least two other reasons based on common brain function. 'Advocates of dual coding theory argue that people retain information best when it is encoded in both visual and verbal codes' (Byrnes, 2001, p. 51). Therefore, we surmise that using visualizations improves retention of the mathematics taught. Secondly, we find that ' . . . after studying pictures along with the words, participants . . . easily reject items that do not contain the distinctive pictorial information they [brains] are seeking' (Schacter, 2001, p. 103). Schacter's research implies that our students may ignore symbolic work if not accompanied by visualizations. Fortunately, we have hand-held devices that can quickly produce most of the visualizations needed in school mathematics at the levels of calculus and below. The author's work with in-service mathematics teachers revealed that most teachers use visualizations to confirm pencil and paper procedures. Is this pedagogy sufficient to produce better recall and reduce rejection?

There is another consideration regarding the use of visualizations, and it is the timing of its use. 'Any attempt to reduce transience [memory loss over time] should try to seize control of what happens in the early moments of memory formation, when encoding processes powerfully influence the fate of the new memory' (Schacter, 2001, p. 34). In addition ' . . . because we have visual, novelty-loving brains, we're entranced by electronic media' (Ackerman, 2004, p. 157). Therefore, in creating a better memory of the mathematics we are teaching, use a hand-held (or other) electronic device at the beginning of a lesson. This draws attention to the mathematics. Once we have student's attention, the visualization will be the key to positively influencing the very existence of the memory.

## Meaning

Mathematics educators would probably argue that we should provide meaning to the rather abstract mathematical concepts we teach. The reasons for adding meaning to the mathematics we teach likely varies from teacher to teacher. The methods for attaching meaning may vary as well. From a neuroscientist view, Steven Pinker offers an idea: 'The human mind, we see, is not equipped with an evolutionarily frivolous faculty for doing Western science, mathematics, chess, or other diversions. . . . The mind couches abstract concepts in concrete terms' (1997, pp. 352–53). If the brain attempts to understand abstract ideas by interpreting them in concrete terms, this suggests that teachers can help the brain understand abstractions by providing meaning through real-world contexts that make sense to our students. A real-world context should be simple, familiar (or easily explainable), and lead directly to the mathematics we are teaching. Supplemental to enhancing understanding, real-world context will add the emotional connection (tag) that will improve memory.

However, there is more to the idea of adding meaning through real-world contexts.

> When a child has a personal stake in the task, he can reason about that issue at a higher level than other issues where there isn't the personal stake. . . . These emotional stakes enable us all to understand certain concepts more quickly. (Greenspan and Shanker, 2004, pp. 241–42)

In the process of adding meaning through familiar contextual situations, we benefit from our students being able to function at a higher cognitive level and understand concepts/procedures more quickly. The simple and familiar context Greenspan and Shanker used in one study was to manipulate candies in the process of teaching addition. The point is that the emotional, or personal stakes, can be extremely simple. Further, the researchers suggest that the contextual situation be used to **teach** mathematics. There is no mention of using applications, something we use **after** the mathematics has been taught.

Finally, and repeated for emphasis, we find: 'A semantic level of processing, which is directed at the meaning aspects of events, produces substantially better memory for events than a structural or surface level of processing' (Thompson and Madigan, 2005, p. 33).

## Conclusion

Neuroscience has other research results on basic brain function that we can apply to education. Memory considerations, an enriched teaching/learning environment, attention, and accessing unconscious processing come to mind. The issue is whether thinking of teaching as being about 'explaining' a list of topics followed by practice can be the paradigm to facilitate a brain-based pedagogy. Technology (hand-held and other) plays a significant role in facilitating the use of connections, pattern generalization, visualizations, and meaning.

The author proposes that algebra be taught with function as an underlying theme, which is taught with the daily use of graphing technology and TI-Navigator. For more information about the use of function as a central theme, see articles posted at www.math.ohio-state.edu/~elaughba/.

## References

Ackerman, D. (2004). *An Alchemy of Mind: The Marvel and Mystery of the Brain.* New York: Scribner.

Bagley, S. (2008). *Train Your Mind: Change Your Brain.* New York: Ballantine Books.

Brodie, R. (1996). *Virus of the Mind: The New Science of the Meme.* Carlsbad, CA: Hay House.

Burton, R. A. (2008). *On Being Certain: Believing You Are Right Even When You're Not.* New York: St Martin's Press.

Byrnes, J. P., (2001). *Minds, Brains and Learning: Understanding the Psychological and Educational Relevance of Neuroscientific Research.* New York: The Guilford Press..

Devlin, K. (2000). *The Math Gene: How Mathematical Thinking Evolved and Why Numbers Are Like Gossip.* New York: Basic Books.

Edelman, G. M. (2006). *Second Nature: Brain Science and Human Knowledge.* New Haven, CT: Yale University Press.

Greenspan, S. I. and Shanker, S. G. (2004). *The First Idea: How Symbols, Language, and Intelligence Evolved from our Primate Ancestors to Modern Humans.* Cambridge, MA: Da Capo Press.

Hawkins, J. (2004). *On Intelligence.* New York: Times Books.

Lehrer, J. (2009). *How We Decide.* New York: Houghton Mifflin Harcourt.

Pinker, S. (1997). *How the Mind Works.* New York: W. W. Norton & Company.
Restak, R. (2006). *The Naked Brain.* New York: Three Rivers Press.
Schacter, D. L. (2001). *The Seven Sins of Memory: How the Mind Forgets and Remembers.* Boston, MA: Houghton Mifflin Company.
Thompson, R. F. and Madigan, S. A. (2005). *Memory.* Washington, DC: Joseph Henry Press.

Chapter 2

# ICT in the United States: Where We Are Today and a Possibility for Tomorrow

Gail Burrill
*Michigan State University, US*

*Every school mathematics program should provide students and teachers with access to tools of instructional technology, including appropriate calculators, computers with mathematical software, Internet connectivity, handheld data-collection devices, and sensing probes.*

(NCTM, 2003)

## The Present

### ITC in Schools

Technology in a variety of forms is becoming increasing common in classrooms in the United States, as it is in much of the rest of the world. In fall 2005, nearly 100 per cent of public schools in the United States had access to the internet, compared with 35 per cent in 1994; 94 per cent of public school instructional rooms had internet access, with little difference by school characteristics (Wells and Lewis, 2006). Interactive white boards are becoming more and more prevalent in schools; in 2007, 21 per cent of the classrooms were equipped with them, 35 per cent in 2009, and the prediction is that by 2011 nearly 50 per cent of US classrooms will have interactive white boards (ICT Products Market Report, 2008). Another type of technology, classroom response systems or 'clickers', was used in 16 per cent of the classrooms (Texas Instruments Education Technology, 2009).

Online courses are becoming more common, particularly for students in small schools primarily as a way of reaching underserved students, not as a preferable alternative to classroom instruction. Recently, however, policy-level calls are looking for online instruction as a cost-efficiency strategy. As of 2008, 25 states have or support statewide virtual schools, usually for middle or high school grade spans; four states offer elementary programmes. An increasing number of states (currently 27) offer computerized state assessments to at least some students (Bausell, 2008). The state of Maine has programmes in place to provide every student with a laptop or access to equivalent technology, but how this is reflected in classrooms, particularly mathematics classrooms, has not been documented.

## Calculators/Computers

Despite the advances in technology as a communication and interaction medium, the hand-held calculator is still the most universally adopted technology in mathematics classrooms. About 76 per cent of the US 4th grade students have access to a four-function calculator, a number that has been relatively constant over the years, with about 6 per cent having a graphing calculator (NCES, 2008). The use of graphing technology for students in grade 8 (about age 14) shows a slight positive trend (Table 2.1).

The frequency of calculator use at the secondary level varies according to the type of course, with students in algebra and higher courses more likely to use graphing and other calculators than students in pre-algebra or regular courses (Braswell et al., 2001). In 2005 (NCES, 2008), 62 per cent of grade 12 students reported using a calculator for homework every day or almost every day with another 16 per cent using the calculator two or three times a week; 64 per cent of the students usually used a graphing calculator. About 43 per cent of the students used a graphing calculator outside of school. Less than 25 per cent of students in grade 8 and 20 per cent of the students in grade 12 reported doing anything with computers; this includes the few who reported using computer software as a resource for extra support for learning (NCES, 2008).

**Table 2.1** Percents of Eighth-grade Students in NAEP Studies Reporting Use of Graphing Calculator in Mathematics Class

| Year | Always | Sometimes/ not often | Never |
|---|---|---|---|
| Use of graphing calculators in math class | | | |
| 2005 | 19 | 27 | 55 |
| 2007 | 21 | 25 | 49 |
| Use of graphing calculators on tests | | | |
| 2000 | 23 | 45 | 32 |
| 2007 | 20 | 54 | 26 |

Braswell et al. 2001; NCES 2003; NCES 2008

Nearly every mainstream textbook series at the high school level includes graphing calculator activities, and five of the series include *TI-Nspire*™ activities, although in most of these the applications are not in fact necessary for the course (Senk et al., 2004). According to a scrutiny of state assessment administration guides, graphing calculators are recommended or allowed in 40 states and required by 7 states for end of course or high school exit examinations.

At the post-secondary level, according to Dossey et al. (2008) the number of calculus sections at 2-year post-secondary institutions using graphing calculators had increased from 44 per cent of the sections in 1995 to over 75 per cent in 2005, in particular for non-mainstream calculus. At 4-year institutions usage also increased from about 30 per cent of the sections in 1995 to about 50 per cent in 2005. The one downward trend over that same period is that use of computers in mainstream calculus sections for both 2- and 4-year institutions decreased from 35 per cent and 31 per cent respectively to 20 per cent.

**Mathematics Software**

Currently there are many examples of technology applications in mathematics teaching. The No Child Left Behind Act (2002) of the US Congress requires states that want to receive federal funding to develop assessments in basic skills to be given to all students in certain grades. Motivated by this law, schools in the United States are struggling to have their students become 'proficient or better' on their state designed mathematics assessments. Strategies to support those who fail often involve interactive, self-paced software that provide drill in arithmetic skills or mathematical procedures. Many of the programmes are supplementary to the curriculum and require students to spend time during the day on mathematics in addition to their regular mathematics classes but are often implemented with no connection to the school programme or to the mathematics that is being taught in the regular class sessions. According to a US Department of Education analysis, only one programme of this nature, the *I CAN Learn® Education System*, which supports students aged 11–14 in learning the fundamentals of mathematics and algebra, has been shown to be somewhat effective in raising student achievement (What Works Clearinghouse, 2009).

The use of software specifically for teaching and learning as a part of mathematics instruction has remained relatively constant with about 70 per cent of high school teachers reporting no use at all. *The Geometer's Sketchpad*, the most commonly used dynamic geometry software in the United States, was used by approximately 7 per cent of the teachers (Texas Instruments Education Technology, 2009) and *Fathom*, software for learning statistics, is estimated to be used by about 6000 secondary teachers or about 1 per cent of the teachers (Finzer, 2009). Secondary teachers are also beginning to use software such as *GeoGebra*, and sessions on how to do so are becoming standard at mathematics conferences, but the extent of use is difficult to track. None of these data indicate whether and how often teachers use the software, however.

## Classroom Instruction

Overall, the research on what teachers actually do with technology and on its impact on student learning is limited. A state-by-state report on innovation, *Leaders and Laggards* (2009), found little data relating to states' use of technology to rethink the delivery of education or improve outcomes and concluded that, 'Educators often give little thought to how technology might modernize education delivery and thus improve teaching and learning. Schools, for example, frequently purchase computers without clear learning goals – and eventually let them languish at the back of classrooms.' (p. 46). Smith et al. (2005) used the words 'boon or bandwagon' in discussing research on the use of interactive white boards. They concluded that although teachers and students appeared strongly in favour of them, there is insufficient evidence to identify any impact on student learning and achievement. An indication of their popularity, however, is that over 400 teachers lined up an hour early for a session on interactive technology at the 2009 Annual Meeting of the National Council of Teachers of Mathematics; the presentation focused primarily on how to use the technology, not on the mathematics students could learn. This is not unusual in the history of research on technology and learning that typically shows no significant difference due to the mere introduction of technology. The most recent examples include the large IES-funded technology study (Dynarski et al., 2007) and recent studies of one laptop per child and interactive whiteboards. All of this research implies that it is not a reasonable expectation that introducing technology, by itself, will impact learning substantially.

Leatham and Peterson (2005) argue, 'Much of what we know about the use of technology in the teaching and learning of mathematics is anecdotal and might be referred to as "possibility" research.' Some research on the use of hand-held technology and on dynamic software does suggest, however, that such tools can have a positive influence on what and how students learn mathematics. (See e.g., Burrill, et al., 2002; Ellington, 2003, 2006; Graham and Thomas, 2000; Schwarz and Hershkowitz, 1999; Hollar and Norwood, 1999 related to graphing calculators; Jones, 2002; Hollebrands et al., 2008 on dynamic geometry software). The existing research makes clear that what is important is how teachers use the technology. Some large-scale studies are currently being conducted (e.g., Owens et al., 2008) but given the magnitude of the research needed, these are few and scattered in terms of approaches and seem to be neither coherent nor cumulative.

Several recent trends, however, seem promising as ways to enhance learning of core mathematical concepts. The following section describes some new thinking about the use of dynamic interactive software in an applet-based environment.

## New Opportunities

Despite the lack of good information about the impact of technology on classrooms, the technology itself continues to advance at a phenomenal rate. Typically, as described above in terms of textbooks, the use of hand-held devices and computers as part of classroom instruction in the past has primarily been as a toolbox to *perform calculations* and *carry out procedures* and their role has not been integral to the lesson; concepts are developed by the teacher with technology 'added on' for students to check solutions or apply those concepts. Such uses of the technology are important, and have enabled more students to engage in high quality mathematics and to solve problems that they could not have done without it. Dynamic interactive technology, however, has opened up new opportunities to consider how technology can be used from another perspective, as a *tool for learning* by enabling the creation of 'environments' in which students can play with a mathematical idea in a variety of ways but where the opportunity to go astray, both mathematically and operationally, is limited.

By imposing constraints on what is possible, teachers and students actually have more freedom to explore central mathematical concepts in deeper ways. Such environments are similar to using applets, (see e.g., the *Rice University Virtual Lab in Statistics* at http://onlinestatbook.com/rvls.html ; *SimCalc* at www.kaputcenter.umassd.edu/products/software/ ; or the Freudenthal Institute website at www.fi.uu.nl/wisweb/en/) and have certain characteristics:

- Little knowledge about the operation of the hand-held or computer is required to use the set of applets.

- The fundamental idea is simple and straightforward. The development has both mathematical fidelity (is mathematically sound and accurate) and pedagogical fidelity (does not present obstacles that obscure or hide critical features).
- The design is based on an action/consequence/reflection principle, where students take an action on a mathematical object, immediately see the consequences, and reflect on the implications of these consequences for a particular mathematical objective. (Dick et al., 2007)
- The interaction is typically driven by one object such as a point, slider, shape or graph and is not menu driven.
- The action/consequence activity is usually composed of two or three carefully sequenced applets designed to have students investigate a core mathematical concept.
- The applets are intended to support the existing curriculum and can be packaged or sequence to complement a standard chapter or unit of study.

The following examples illustrate how 'action/consequence' applets might be used to develop understanding of fundamental mathematical concepts. The activities stem from a careful examination of high stakes assessments, many developed by each state to comply with No Child Left Behind, usually end of course or school leaving requirements necessary to receive a secondary diploma. From item analyses of these assessments across states (Burrill and Dick, 2006), we were able to identify content topics on which students consistently had low achievement scores. Based on this information and on research about student learning, action/consequence experiences can be designed to help students better understand the core concepts that had emerged as problematic.

As an illustration, the assessment analyses and the research (e.g., Kuth et al., 2007) suggest that students have trouble with the concept of the variable. Students also have poor achievement scores with tasks that involve reasoning about mathematical concepts and with solving systems of equations. The following examples describe action consequence documents that might be used to help develop student understanding of the mathematics involved. Figure 2.1 shows an action/consequence document where the value of an expression is displayed as students move a point on a number line with integer coordinates. As students move the point, they can investigate a variety of questions while developing a sense of variable: How does the output change as the value of $x$ changes? When, if ever, will the output be divisible by 4? Will the output be zero? Why or why not? Can you get the same output for different values of $x$?

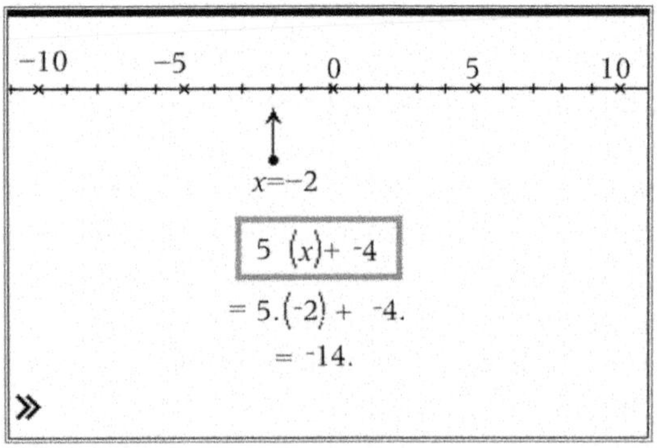


**FIGURE 2.1** An Expression and a Number Line

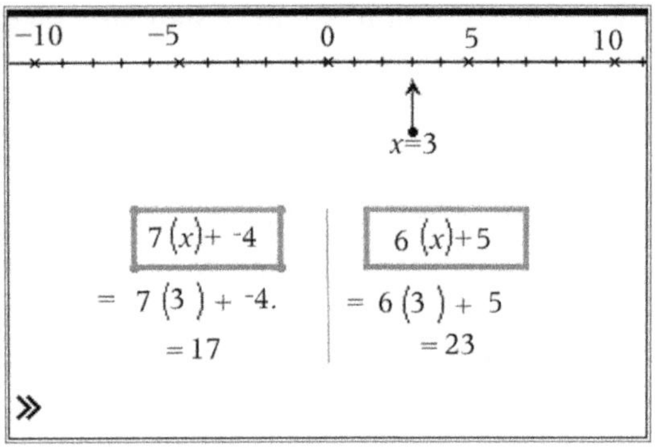


**FIGURE 2.2** Two Expressions and a Number Line

Dragging a point as shown in Figure 2.2 allows students to investigate the difference between expressions and equations, laying the foundation for thinking about an equation as the equality of two expressions, and building on their developing sense of a solution for a more general linear equation in one variable. Probing questions push students to look at these relationships and to reason about the variables, the coefficients and the output.

'What is a Solution?' (Figure 2.3) involves an investigation of a solution to a linear equation in two variables. In the first activity, students drag a point whose coordinates satisfy a 'current' equation until the current equation matches the 'goal' equation, then mark that point. Questions can be posed to drive student thinking about how to extend the concept of solution

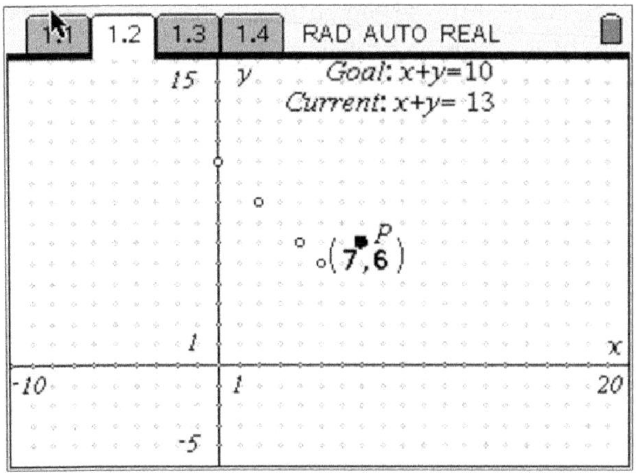


**FIGURE 2.3** What Is a Solution

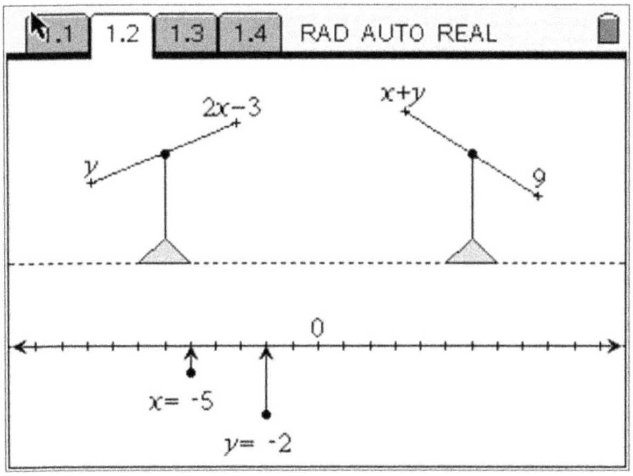


**FIGURE 2.4** Balanced Systems

to a linear equation in two variables. In examining strategies for finding more points that satisfy the target equation, students can also make a connection to slope. Balanced Systems (Figure 2.4) connects to the earlier documents and allows students to experiment with a system of equations by finding the 'balance point' for both equations at the same time. The next figures use the balance scales to investigate dependent and inconsistent systems.

As students make conjectures about strategies for balancing both the equations at the same time, they can develop a sense of how the two equations are interacting – and with the right questions, make connections to the graphs of those equations and what the intersection would mean.

## Conclusion

Technology has many faces in mathematics classrooms. One of the newest involves dynamic interactive software activities, such as those described in this chapter, for either a hand-held or a computer that engages students in thinking about concepts across mathematical domains, using the notion that allowing students to repeatedly take a mathematical action, observe the consequences and reflect on the mathematical implications improves understanding. Currently some activities such as those described above are being developed as a part of a freeware offering called *Math Nspired* (2009) from Texas Instruments Education Technology. While such activities raise questions for mathematics education research, they seem like the next step in taking advantage of technology to scaffold a better understanding of central mathematical concepts for students.

## References


Bausell, C. V. (2008). 'Tracking U. S. trends in Technology Counts 2008 STEM'. *Education Week,* 27 (30), 40–41.

Braswell, J., Lutkus, A., Grigg, W., Santapau, S., Tay-Lim, B. and Johnson, M. (2001). *The Nation's Report Card: Mathematics 2000.* Washington DC: National Center for Education Statistics.

Burrill, G., Allison, J., Breaux, G., Kastberg, S. E., Leatham, K. and Sanchez, W. B. (2002). *Handheld Graphing Technology in Secondary Mathematics: Research Findings and Implications for Classroom Practice.* Dallas, TX: Texas Instruments, Inc.

Burrill, G. and Dick, T. (2006). Presentation at National Council of Supervisors of Mathematics Annual Meeting, Atlanta, GA.

Dick, T., Burrill, G. and Brady Gill, L. (2007). 'New technologies offer new ways to engage students'. *NCSM Newsletter,* 38 (1), 19–21.

Dossey, J. and Halvorsen, K. and McCrone, S. (2008). *Mathematics Education in the United States 2008: A Capsule Summary Fact Book.* Reston, VA: National Council of Teachers of Mathematics.

Dynarski, M., Agodini, R., Heaviside, S., Novak, T., Carey, N., Campuzano, L., Sheila Heaviside, S., Novak, T., Carey, N., Campuzano, L., Means, B., Murphy, R., Penuel, W., Javitz, H., Emery, D. and Sussex, W. (2007). 'Effectiveness of reading and mathematics software products: Findings from the first student cohort'. (Publication No. 2007–4005). Retrieved 18 November 2007, from the Institute of

Education Sciences, U.S. Department of Education Website: ies.ed.gov/ncee/pdf/20074005.pdf

Ellington, A. J. (2003). 'A meta-analysis of the effects of calculators on students' achievement and attitude levels in precollege mathematics classes'. *Journal of Research in Mathematics Education*. 34 (5), 433–64.

—(2006). 'The effects of non-CAS graphing calculators on student achievement and attitude levels in mathematics: A meta-analysis'. *School Science and Mathematics*. 106 (1), 16–26.

Finzer, W. (2009). Personal correspondence.

Graham, A. T. and Thomas, M. O. J. (2000). 'Building a versatile understanding of algebraic variables with a graphic calculator'. *Educational Studies in Mathematics*, 41 (3), 265–82.

Hollar, J. C. and Norwood, K. (1999). 'The effects of a graphing-approach intermediate algebra curriculum on students' understanding of function'. *Journal for Research in Mathematics Education*, 30 (2), 220–26.

Hollebrands, K., Laborde, C., and Strässer, R. (2008). 'Technology and the learning of geometry at the secondary level'. In Heid, K. and Blume, G. (eds), *Research on Technology and the Teaching and Learning of Mathematics*, pp. 155–206. Greenwich, CT: Information Age.

Interactive Displays / ICT Products Market Report, 2008, Decision Tree Consulting.

Jones, K. (2002). 'Implications for the classroom: Research on the use of dynamic software'. *Micromath*.18 (3), 18–21.

Kuth, E., Alibali, M., Stephens, A. and Asquith, P. (2007). 'Middle school mathematics teachers' knowledge of students' understanding of core algebraic concepts: Equal sign and variable'. *Mathematical Thinking and Learning*, 9 (3), 249–72.

*Leaders and Laggards: A State-by-state Report Card on Educational Innovation.* (2009). Center for American Progress, U.S. Chamber of Commerce, and Frederick M. Hess, director of education policy studies at the American Enterprise Institute. www.uschamber.com/assets/icw/09reportcard/09_leadersandlaggards.pdf (accessed on 9 November 2009).

Leatham, K., and Peterson, B. (2005). 'Research on teaching and learning mathematics with technology: Where do we go from here?' In Lloyd, G. M., Wilson, M., Wilkins, J. L. M., and Behm, S. L. (eds) Proceedings of the 27th annual meeting of the North American Chapter of the International Group for the Psychology of Mathematics Education. Virginia Technology Institute. [CD-ROM]. Eugene, OR: All Academic.

National Center for Education Statistics (NCES) (2003). *Digest of Educational Statistics: 2002.* Washington DC: U.S. Department of Education.

National Center for Education Statistics (NCES) (2008). *National Assessment of Educational Progress (2005).* Data Explorer High School Transcript Study.

National Council of Teachers of Mathematics. (2003). *The Use of Technology in the Teaching and Learning of Mathematics.* NCTM Position Statement. Reston, VA: Author. www.nctm.org/uploadedFiles/About_NCTM/Position_Statements/technology.pdf (accessed on 9 November 2009).

No Child Left Behind Act of 2001. (Pub.L. 107–110, 115 Stat. 1425, enacted 8 January 2002). Washington DC: United States Congress.

Owens, D. T., Pape, S. L., Irving, K. E., Sanalan, V., Boscardin, C. K. and Abrahamson, L. (2008). 'The connected algebra classroom: A randomized control trial', *Proceedings for Topic Study Group 22, Eleventh International Congress on Mathematics Education*. Monterrey, Mexico, Retrieved 2 July 2009 from tsg.icme11.org/document/get/249

Schwarz, B. B., and Hershkowitz, R. (1999). 'Prototypes: Brakes or levers in learning the function concept? The role of computer tools'. *Journal for Research in Mathematics Education*, 30, 362–89.

Senk, S., Burrill, G. and Olanoff, D. (2004). 'Knowledge for teaching algebra: What can we learn from examining textbooks?' Paper prepared for the Association of Mathematics Teacher Educators Annual Meeting, San Diego, CA.

Smith, H., Higgins, S., Wall, K. and Miller, J. (2005). 'Interactive whiteboards: Boon or bandwagon? A critical review of the literature'. *Journal of Computer Assisted Learning*. Blackwell Publishing Ltd, 21, 91–101

Texas Instruments Education Technology. (2009). Marketing Survey.

Texas Instruments Education Technology. (2009). Math Nspired. www.ti-mathnspired.com/ (accessed on 9 November 2009).

Wells, J., and Lewis, L. (2006). *Internet Access in U.S. Public Schools and Classrooms: 1994–2005* (NCES 2007–020). U.S. Department of Education. Washington, DC: National Center for Education Statistics.

What Works Clearing House (2009). Institute for Education Sciences. U.S, Department of Education. http://ies.ed.gov/ncee/wwc/ (accessed on 9 November 2009).

Chapter 3

# ICT in France: Development of Usages, Institutional Hesitations and Research Questions

Luc Trouche and Ghislaine Gueudet
*INRP, Lyon, France*

## Introduction

In France, as in other countries, it is acknowledged that tools evolve faster than their usage, particularly in school mathematics, despite institutional pressure (Trouche, 2005). However, evidence of deep changes is appearing, fostered by the internet and hand-held technologies. For example, teachers design and share resources on a large scale and innovative assessment procedures and teacher training devices are promoted by the French education ministry (Mission e-Éduc, 2008). Portraying a comprehensive picture of such an evolving landscape is difficult; therefore, we present here three snapshots to give an idea of the main current trends, and consider the development of new theoretical tools required to analyse these new phenomena.

(1) Collaborative design of online resources: the *Sesamath* association

A major development in France is the growth of *teachers' online associations*, which enable the design and sharing of teaching resources, in mathematics in particular. The *Sesamath*[1] association (Montessinos and Kuntz, 2008) has only around 70 members, almost all of them secondary school teachers, but its website reports more than 1.3 million hits each month. The objectives of *Sesamath* are to promote Information and Communications Technology (ICT) integration, to foster teachers' collaborative work and to support students' learning. The resources freely available from the association comprise:

- an e-exercise module, *Mathenpoche*, which covers the whole curriculum from grades 6 to 9 with interactive exercises. Mathenpoche is probably the main reason for *Sesamath*'s success (Bueno-Ravel and Gueudet, 2009): its use is very simple, and no similar e-exercises resource exists in France;

- a set of lessons and exepreadsheet (*Casenpoche*), a dynamic geometry software (*Tracenpoche*), srcises in *Acrobat Reader* document (pdf) or *OpenOffice* format (called *Mathadoc*);
- a range of software: a simulation of geometry tools (*Instrumenpoche*);
- online textbooks, with traditional lessons and exercises complemented by interactive resources, for example, links towards interactive geometry figures; these textbooks have counterparts on paper (much cheaper than the ordinary textbooks).

*Sesamath* was established in 2001, by a group of approximately 15 teachers who intended to share online resources each of them had designed. They soon started collectively developing *Mathenpoche*, which was completed in 2006: subgroups were responsible for particular themes. During this period, other members joined, new projects emerged; collaborative work started around the textbooks (Gueudet andTrouche, 2009a). A special website (*Sesaprof*) was opened in 2008 for *Sesamath* resources users. On this site, users could share their experiences and more than 7000 teachers have already registered. The association's present projects comprise, in particular, a resource called *LaboMep*, incorporating a new version of *Mathenpoche* together with other software, which can be easily integrated in a Virtual Learning Environment and can also be translated into different languages. Moreover, collaborations started with researchers, in order to improve the quality of the resources and of their uses.

*Sesamath* has sought support from the Ministry of Education, but they have always declined, citing that designing resources is not a part of teachers' professional roles. The Ministry's actions to support ICT integration takes other forms, in particular the development of new modes of assessment (§ 2), and the creation of new training programmes (§ 3).

(2) New assessment procedures, the *mathematics practical test* at the French secondary school final examination (baccalauréat)

It is well known that assessment both raises awareness of the need for and serves to pilot curricular development. Crucially, curricular development has raised the awareness in educational institutions of the need for ICT integration. From this point of view, the case of the mathematics practical test within the French baccalaureate is very interesting.

Since 2007, French high schools piloted a new test (end of 10th grade), the objective of which was 'to evaluate students' competencies in using calculators and some software designed for mathematics [ . . . ] to solve a given problem' (MEN, 2007). This new test has deeply (Figure 3.1) modified the type of question posed, the nature of the exam and the environment in which it is sat (ICT explicitly solicited, open-ended questions, examiner has an active role).

Such a 'revolution' has provoked a lot of discussion, protests[2], but, little by little, teachers are changing the ways in which they prepare their students for their final examination. In an official report, a national inspector (Fort, 2007) writes: 'this new test forges new relationships between students and mathematics [ . . . ], it allows students to experiment and check through using ICT [ . . . ]; it encourages new teaching approaches, giving a more prominent role to investigative processes'. This report concludes:

> the wider adoption of this test, which exists within the frame of the current curriculum, will force mathematics teaching to develop more coherency with its purpose: how could mathematics, with the tools currently available, solve problems, develop experimentation, and the motivation for research?

Unfortunately, wider adoption will not occur until 2013 because it is considered better to wait for a change of the final examination as a whole, rather than to change one aspect.

Let $f$ be the function defined on R by: $f(x) = -x + \sqrt{x^2 + 4}$, its graph $C$, $\alpha$ a real number, $M$ and $N$ the points on $C$ with abscissae $\alpha$ and $-\alpha$.

1) Graph $C$ with your chosen software. Call an examiner to check your graph.
2) Try several values of $\alpha$ and make a conjecture about the locus of the point $I$ (the intersection of the tangents to $C$ through points $M$ and $N$). Call an examiner to check your graph.
3) Determine, as a function of $\alpha$, the coordinates of M and N. Justify your previous conjectures.

Required outcomes:
- visualization of the locus of point $I$
- justified answers to question 2.

**FIGURE 3.1** Example of practical test

(3) Emerging teachers training devices grounded in collective work

*Pairform@nce*[3] was set up in 2006 by the Ministry of Education in France after a successful experiment in Germany (Intel Lehren-Aufbaukurs online). It is an in-service teacher training project, aimed at supporting ICT integration, for any level of class and school domain. In this project, designers devise *training paths,* which will be used by teacher trainers to develop their own training programmes. The training programmes must comply with certain principles: blended training using a shared platform; collective preparation of classroom sessions integrating ICT tools, and a succession of seven stages – introduction of the training, selection of themes and constitution of teams, co-and self-training, design of classroom situation, implementation in class, reflective analysis and evaluation.

Thus, from a research point of view, there are many questions that can be explored about this project, particularly: Can this kind of training modify teachers' practices? Can teacher trainers use training paths they did not design themselves?

Our research and design team (supported by INRP-National Institute for Pedagogical Research) designed in particular, for the training of secondary schools teachers (Gueudet et al., 2008), two paths in mathematics: 'personalisation using e-exercises' and 'inquiry-based teaching with dynamic geometry software'. These paths were simultaneously designed and tested in 2007–2008; we recorded significant evolutions of the trainees' collective work, they all produced technology-rich lessons, and they also contributed to the design of the paths by their comments and remarks about the experimental training. In 2008–2009, these paths were cross-experimented (Artigue, 2007): the designers of one path became trainers for the other. The familiarization process of the trainers required a significant amount of work, and led to important developments: they selected from the suggested resources, adapted them, added resources, for example, trainers more acquainted with e-learning added resources for distance learning, and so on. The trainers contributed to further improving the design of the paths. Our team designed tools for communication between trainers and path designers (a *trainer logbook* in particular, Mialles-Viard Metz et al. 2009).

In *Pairform@nce,* like in *Sesamath,* ICT is associated with collective aspects. While the collaboration in *Sesamath* was spontaneous, in *Pairform@nce* it was required by the institution, with an aim of developing ICT integration.

Despite this essential difference, similarities exist: appropriate tools are required to support collaboration; the resources must stay alive, integrating the users' experience.

(4) New needs for researchers: understanding *documentational* and *professional geneses*

Teachers' resources, of course, have always been alive, evolving alongside teachers' practices. However, this fact is enlightened and intensified by the wider usage of online resources. Teachers interact with a lot of resources, throughout their work, both in and outside of school. It naturally leads to a more comprehensive notion of what a resource is, to 'think of a resource as the verb re-source, to source again or differently' (Adler, 2000, p. 207), including everything that teachers are likely to encounter in their work: ideas of problems, discussions between teachers, orally or online; students' sheets, and so on.

## Conclusion

Following this line, we have proposed a new perspective for the study of teachers' professional evolution, where the researcher's attention is focused on the resources, their *appropriation* and *transformation* by the teacher or by a group of teachers working together (Gueudet and Trouche, 2009b). In this situation, we distinguish between *resources*, available for teacher's work, and *documents*, results of this appropriation process. Thus a document is much more than a list of resources; it is *saturated* with the teacher's experience, like a word, for a given person, is saturated with sense in a Vygotskian perspective. A document is an encapsulation of resources, professional experience and knowledge. In this perspective, there is a strong interaction between *documentational geneses* and *professional geneses*.

It raises sensitive methodological issues: observing long-lasting phenomena and processes, set in different places in and outside of school, particularly in the teacher's own home, individual processes as well as collective ones, is a challenge. We nevertheless consider it fundamental to develop this perspective, if, considering the ocean of contemporary mathematics education, we want to distinguish between surface waves and deeper undercurrents.

## Notes

1 www.sesamath.net/
2 See, for example, the Educmath forum: http://educmath.inrp.fr/Educmath/en-debat/epreuve-pratique/
3 www.pairformance.education.fr

## References

Adler, J. (2000). 'Conceptualising resources as a theme for teacher education'. *Journal of Mathematics Teacher Education* 3, 205–24.

Artigue, M. (2007). 'Digital technologies: A window on theoretical issues in mathematics education'. In Pitta-Pantazi, D.and Philippou, G. (eds) Proceedings of CERME 5, pp. 68–82, Larnaca, Cyprus, http://ermeweb.free.fr/CERME5b (accessed on 6 September 2009).

Bueno-Ravel, L. and Gueudet, G. (2009). 'Online resources in mathematics: Teachers' geneses and didactical techniques'. *International Journal of Computers for Mathematical Learning*, 14 (1), 1–20.

Fort, M. (2007). *Rapport sur l'expérimentation d'une épreuve pratique de mathématiques au bac S, MEN*, on line http://educmath.inrp.fr/Educmath/en-debat/epreuve-pratique/rapportep (accessed on 6 September 2009).

Gueudet, G., Soury-Lavergne, S. and Trouche, L. (2008). 'Soutenir l'intégration des TICE: quels assistants méthodologiques pour le développement de la documentation collective des professeurs?' Exemples du SFoDEM et de Pairform@nce. *Communication au colloque DIDIREM*, Paris, septembre 2008.

Gueudet, G. and Trouche, L. (2009a). 'Conception et usages de ressources pour et par les professeurs : développement associatif et développement professionnel'. *Dossiers de l'ingénierie éducative* 65, 78–82.

Gueudet, G. and Trouche, L. (2009b). 'Towards new documentation systems for mathematics teachers?' *Educational Studies in Mathematics* 71 (3), 199–218.

Mailles-Viard Metz, S., Soury-Lavergne, S. and Daubias, P. (2009). 'Tool for supporting appropriation of on-line teacher training'. In A. Méndez-Vilas, A. Solano Martín, J. Mesa González, J. A. Mesa González (eds) *Research, Reflections and Innovations in Integrated ICT in Education*, Formatex, pp. 845–48. ACTI. halshs-00376559, on line www.formatex.org/micte2009/book/845–848.pdf (accessed on 6 September 2009).

MEN (Ministère de l'Education Nationale) (2007). *L'épreuve pratique du baccalauréat de la série scientifique*, on line http://eduscol.education.fr/D1115/epr_pratique_presentation.htm (accessed on 6 September 2009).

Mission e-Éduc (2008). Pour le développement du numérique à l'école, Rapport de la mission e-Éduc, on line http://media.education.gouv.fr/file/2008/24/5/Pour_le_developpement_du_numerique_a_l_ecole_27245.pdf (accessed on 6 September 2009).

Montessinos, B. and Kuntz, G. (2008). 'Sesamath: A teachers society, to create and to spread mathematical resources through the Internet'. *Proceedings of the*

*Eleventh International Congress on Mathematical Education* , Monterrey, Mexico. 2008

Trouche, L. (2005). 'Calculators in mathematics education: A rapid evolution of tools, with differential effects'. In D. Guin, K. Ruthven, L. Trouche (eds), *The Didactical Challenge of Symbolic Calculators: Turning a Computational Device into a Mathematical Instrument*, pp. 9–39, New York: Springer.

Chapter 4

# ICT and the English Mathematics Curriculum

Sue Pope
*Programme Manager, Mathematics, Qualifications and Curriculum Development Agency, UK*

## Introduction

The use of technology has always been explicit in the statutory curriculum for mathematics since the first version in 1988. This included the use of calculators, *Logo* and specially developed software. Since then spreadsheets, graphing packages and dynamic geometry have become much more widely available. The potential for using Information and Communications Technology (ICT) to support teaching and learning in mathematics has never been greater. However, as Ofsted (2008) has found, whereas in the early days of computer technology in schools mathematics teachers tended to be in the vanguard, nowadays, students have far less hands on experience during mathematics lessons. Many teachers now use interactive whiteboards – but these are largely for presentation purposes.

> Several years ago, inspection evidence showed that most pupils had some opportunities to use ICT as a tool to solve or explore mathematical problems. This is no longer the case; mathematics makes a relatively limited contribution to developing pupils' ICT skills. Moreover, despite technological advances, the potential of ICT to enhance the learning of mathematics is too rarely realised. (Ofsted, 2008)

There are six major opportunities for learners to benefit from the use of ICT in mathematics:

- learning from feedback
- observing patterns

- seeing connections
- developing visual imagery
- exploring data
- 'teaching' the computer

Examples for realizing some of these opportunities are included in QCA's recent publication *Engaging Mathematics for All Learners.* This includes a range of ways in which teachers are implementing the new secondary curriculum for mathematics. Examples include:

- the use of digital photographs from situations outside the mathematics classroom (including PE lessons and a visit to a local playground) as backgrounds in dynamic geometry software and geometric objects and graphs can be superimposed;
- the use of databases to explore hypotheses in other subjects for example, the Eyam plague in history;
- the use of spreadsheets to investigate the nutritional content of different meals;
- the use of *Logo* to explore loci;
- the use of ICT to allow learners to generate many examples and learn from feedback.

## The New Secondary Curriculum

In 2008 the English secondary curriculum was revised. The curriculum aims of 'successful learners, confident individuals and responsible citizens' were introduced and all the subject programmes of study were represented, placing greater emphasis on key concepts and key processes and reducing prescription and content detail.

Explicit reference to the use of technology in the programmes of study for Key Stage 3 and Key Stage 4 include:

Key concepts (Competence):

1.1c selecting appropriate mathematical tools and methods, including ICT

Key processes (Representing, Analysing, Interpreting, Evaluating):

2.1b compare and evaluate representations and choose between them
2.2a make connections in mathematics
2.2c visualize and work with dynamic images

2.2f explore the effects of varying values and look for invariance and covariance
2.2k make accurate mathematical diagrams, graphs and constructions on paper and on screen
2.2l calculate accurately, selecting mental methods or calculating devices as appropriate
2.2n use accurate notation, including correct syntax when using ICT
2.3d look at data to find patterns and exceptions
2.4a use a range of forms to communicate findings to different audiences

Curriculum opportunities

4g Become familiar with a range of resources, including ICT, so that they can select appropriately

If students are to make informed choices about resources, they need to have used them and understood their potential. This means that they need hands on experience with a wide range of ICT tools to explore mathematics for its own sake and to apply mathematics in a range of contexts. ICT allows students to experience, and explore for themselves, multiple representations. Appreciating the relationships between numerical, graphical and symbolic representations is a key factor in students' success in school mathematics. Dynamic geometry extends this to include geometric representations.

## The Primary Curriculum and Possible Revisions

The national curriculum has always advocated mental methods of calculation as a first resort and the efficient use of calculators. However it is not uncommon to see restricted access to calculators in primary school. Children may have to 'master written methods' before they're allowed to use a calculator. When children do use a calculator, it is likely to be to 'check their work' rather than to explore an aspect of number or operations. Children rarely learn how to use a calculator efficiently, exploiting its functionality, for example, the use of the memory key or to repeat an arithmetic process many times.

There is compelling evidence from the national curriculum tests that many children do not know how to use calculators. The national tests at the end of primary school comprise a calculator and non-calculator paper. Many children attempt to use written methods when tackling questions on the calculator paper.

During 2009 the DCSF consulted on proposed changes to the primary curriculum. It was proposed that the aims for the primary curriculum should be the same as those for secondary: 'successful learners, confident individuals and responsible citizens'. The 'Essentials for learning and life' were proposed and these include ICT capability. ICT has been strengthened throughout the primary curriculum proposals to encourage its integration across the curriculum.

Rather than KS1 and KS2 the progression through the primary curriculum was articulated as early, middle and later stages. Examples from the proposals for 'mathematical understanding' include:

- Early stage – use of calculators to explore number patterns
- Later stage – use ICT to represent number patterns as graphs and using simple formulae
- Later stage – use spreadsheets to model financial situations
- Later stage – use a calculator efficiently as one of a range of strategies for calculation
- Geometry – use ICT to generate instructions for movement, generate and explore geometric patterns and problems
- Statistics – use ICT to store, structure and analyse data, that has been collected for a purpose, to explore possible relationships and interpret findings

These proposals are informed by research and best practice. There is considerable evidence that using calculators from an early age as an exploratory tool can enhance understanding of number and operations.

## ICT and Assessment

While the curriculum expectations around the use of ICT are clear, one of the key factors that impacts on how the curriculum is implemented is assessment. Teacher assessment (TA) is expected at the end of key stages and should be based on the rich entitlement set out in the national curriculum. Alongside TA there has been national testing at the end of KS 1, 2 and 3 since the early 1990s and public examinations at the end of KS4. At KS1 the tests are non-calculator and at all other key stages comprise both calculator and non-calculator components.

In October 2008 the government announced the end of KS3 national tests. This represents an exciting opportunity to reinvigorate TA and ensure that it does indeed draw on a rich evidence base which includes the use of ICT. Since 2008, Assessing Pupils' Progress (APP) has been introduced to schools as part of the DCSF's *Assessment for Learning Strategy*. APP is intended to revitalize TA across both primary and secondary. APP introduces periodic assessment which involves teachers standing back and making a holistic judgement about their learners' achievements and planning next steps and adjustments to their teaching. Teachers are encouraged to draw on a rich evidence base which may include tests and 'assessment tasks' but is not solely reliant on them. Periodic assessment should take place, at most, two or three times a year and teachers are advised to start with a small sample of learners. The APP criteria are based on the national curriculum level descriptions. For more about APP see the QCDA curriculum website.

Using ICT in formal assessments, such as national curriculum tests and general qualifications (e.g., GCSEs, GCEs) may be a long-term goal but is certainly unlikely in the medium term. The logistics of learners having access to ICT, other than calculators, in timed written examinations is not currently practicable. Until 2007 students had to complete course work as part of their GCSE; this could include the use of ICT in an extended piece of work. From 2010, GCSE mathematics will have 25–50 per cent non-calculator assessment, a revision from the current 50 per cent. In addition, the GCSE will include a greater proportion of application and problem solving (from 20% to 50%) to reflect the priorities of the new curriculum. This may provide a greater impetus for teachers to ensure their students do know how to use a calculator (or graphing calculator) effectively.

For post-16 learners as well as GCE, AS and A level mathematics there are free standing mathematics qualifications (FSMQs). FSMQs are available at NVQ levels 1, 2 and 3 (i.e., GCSE grade G–D, GCSE grade C+, GCE respectively) and are intended to support learners' other areas of study. They focus on developing understanding through application. Level 3 FSMQs can be combined to gain an AS use of mathematics. The possibility of a full A level in use of mathematics is currently under consideration. This new A level would include the assessment of extended pieces of work based on using mathematics for modelling and problem solving. The use of ICT is integral to FSMQs and graphing calculators are expected in the assessment of all level 3 FSMQs. For more than ten years there has been a non-calculator paper in GCE mathematics, the removal of this requirement was recently proposed – and there are many arguments for and against this proposal.

## Practical Considerations

Providing students with hands on access to ICT during normal mathematics lesson is a challenge for many schools. These are some of the ways that schools make ICT available:

- a mathematics classroom with a number of networked PCs around the edge of theroom in which small groups of students access a range of software
- a set of laptops for use by pairs of students
- a set of calculators, graphing calculators or other hand-held devices that students use alone or in pairs to explore sets of data, plot graphs, and so on
- a wireless network linking a set of hand-held devices to the teacher's PC.

Whatever choices are made by schools, teachers and departments, the use of technology is likely to remain a key element of the required curriculum.

## References

DCSF. (2008). *Assessment for Learning Strategy.* http://www.education.gov.uk/publications/eOrderingDownload/DCSF-00341-2008.pdf (accessed on 11 July 2009).

Ofsted. (2008). *Mathematics: Understanding the Score.* http://www.ofsted.gov.uk/Ofsted-home/Publications-and-research/Browse-all-by/Documents-by-type/Thematic-reports/Mathematics-understanding-the-score (accessed on 11 July 2009).

QCA. (2007). *National Curriculum.* www.curriculum.qcda.gov.uk (accessed on 11 July 2009).

—(2009). Engaging Mathematics for All Learners. www.qcda.gov.uk (accessed on 11 July 2009).

# Part Two

# What Does Research Tell Us?

In this section, authors write about the role that Information and Communications Technology (ICT) has to play in the learning of mathematics, drawing on their own and others' research. Keith Jones considers some of the issues surrounding using ICT to support teaching geometry. Looking back at the hopes and expectations of earlier writers, he reflects on some of the reasons that technological progress may not have materialized in the classroom to the extent anticipated. Don Passey reports on a longitudinal study carried out in the United Kingdom, analysing the impact that interactive whiteboards can have on the learning and teaching of mathematics. Many aspects of the ways in which cognitive and motivational processes are supported are considered. Walter Stroup, Lupita Carmona and Sarah M. Davis then present their research into the impact of using 'Network-Supported Function-Based Algebra' (NFBA) for supporting the teaching of a standard algebra course in the United States. This use of graphical calculators is offered as an example of a 'forward looking' intervention to help prevent underperformance rather than a 'backward looking' remediation or corrective strategy.

All three provide further evidence that technology can undoubtedly be used to improve engagement and understanding – But what structures need to be in place, both nationally and locally, to support all teachers to harness this potential?

Chapter 5

# The Value of Learning Geometry with ICT: Lessons from Innovative Educational Research

Keith Jones
*University of Southampton, UK*

## Introduction

The 40th anniversary of the launch of the *British Journal of Educational Technology* (BJET) took place in 2009. Also in 2009, coincidentally, ATM instigated MT*i* (*Mathematics Teaching* interactive), an online web-based publication. The first issue of BJET (in 1970) was most concerned with the availability of audio-visual resources (such as 16mm film projectors), whereas the first issue of MT*i* contained, among other things, an article which examines questions for which the answer is 'It's a parabola' (see Figure 5.1 for an example).

**FIGURE 5.1** Sand Dune in Morocco (photo by Fabio Cologna)

Initiatives such as MT*i* are symptomatic of the changes in technology that have taken place over the period since the launch of BJET – computing power has become more ubiquitous, technologies have converged, and the emphasis is increasingly on interactivity that exploits learners' visual and spatial intuition. All this means that it is instructive to review the value of learning school geometry with ICT, given that geometry is both a key component of mathematical theory, *and* the quintessential visual and spatial element (Jones, 2002a).

In what follows, the research selected focuses on learners' use of interactive geometry software, the design of suitable teaching and learning activities, and the nature of relevant teacher professional development. The central theme of the paper is that while ICT has considerable potential in enlivening the teaching and learning of school geometry, there is much to take account of in terms of enabling this potential to be fully realized.

The visual and spatial interactivity offered by ICT means that dynamic geometry software (DGS; examples include *Cabri*, *GeoGebra*, *Sketchpad*, etc.) has become perhaps the best researched area in mathematics education (for a review, see Jones, 2002b). At first glance, dynamic geometry software (DGS) is nothing more than a drawing package. Yet a DGS is more than a simple program for graphics because the user can not only specify geometric relations between initial objects (such as points and lines) but can also grasp such objects and drag them. In a classroom research project (Jones 2000), a class of 12-year-old pupils was studied as they completed a module of work on the topic of the properties of quadrilaterals. The aim of the study was to document the meanings that pupils gained of deductive reasoning through experience with DGS software. It was anticipated that their meaning would likely be shaped, not only by the tasks they tackled and their interactions with their teacher and with other pupils, but also by features of the software.

The evidence from the research study indicated that while using DGS does provide learners with a way of working with geometrical theorems, this is mediated by features of the software, especially in the vital early and intermediate stages of using the software. The research illustrates that even with carefully designed tasks, sensitive teacher input and a classroom environment that encourages conjecturing and a focus on mathematical explanation, it can take quite some time for the benefits of using DGS to emerge. For example, learners take time to understand not only the uses that can be made of the facility to drag on-screen objects, but also what is entailed in constructing an on-screen figure in a way that fully utilizes relevant geometrical theory. Not only that, but a particular issue is whether the opportunities offered with DGS to 'see' mathematical properties with such on-screen support might reduce or even replace any need for mathematical proof – or, on the contrary, whether new ways of promoting learners' understanding of the need for, and the roles of, proof might open up (Hoyles and Jones, 1998).

The issue of time also emerges when one considers what is involved in designing teaching and learning activities for geometry. Research on designing teaching scenarios based on various forms of geometry software, and of integrating them in the regular course of classroom teaching, shows that it can take quite a long time to reach the point where tasks genuinely take advantage of the computer environment (Brown et al., 2003; Christou et al., 2006, 2007; Zachariades et al.; 2007). Such research indicates that geometry tasks selected for use in the classroom should, as far as possible, be chosen to be useful, interesting and/or surprising to pupils. In addition, it can be helpful if classroom tasks expect pupils to explain, justify or reason, and be critical of their own, and their peers' explanations. In particular, the generating of data or the use of measurements, while playing important parts in mathematics, and sometimes assisting with the building of conjectures, are probably best not as an end point to pupils' mathematical activity. Indeed, where sensible and in order to build geometric reasoning and discourage over-reliance on empirical verification, classroom tasks might use contexts or approaches where measurements or other forms of data are *not* generated. In addition to taking time to reach the point where tasks genuinely take advantage of ICT, the issue of finding how to manage classroom time well during actual teaching is also something that research shows has to be worked on.

An especially interesting conundrum relates to using ICT for 3D geometry (Christou et al., 2006; Jones et al., 2009b). It may seem, at first sight, rather odd to be working in 3D geometry on a flat 2D computer screen. Not only that, but the issue of representing 3D objects on a flat screen means that a number of design decisions, unique to 3D software, need to be made by software developers – one being the key decision of how the opening software screen both orients the user to 3D space, and provides a framework for the creation of 3D figures and structures. Inevitably, this has been tackled in different ways by different software developers, yet what learners make of such differences is currently under-researched. What is raised is the issue of just how 'direct' is what is often called 'direct interaction' when using ICT for geometry. As digital technologies develop, it is unclear if learners do feel that they are interacting 'directly' with geometrical theory; what learners may experience is rapidly moving dynamic on-screen images that seem more like computer-generated imagery. The question remains about how the learning of geometry can be facilitated through different digital technologies in a way which successfully builds upon the visual intuition that all of us require in order to understand our experience of physical and mathematical space.

The need to understand learners' use of interactive geometry software, and the need to be able to design suitable teaching and learning activities, point to the importance of research on forms of suitable teacher professional development. In a project involving experienced teachers collaborating in developing ways of providing professional development and support for other teachers (Jones et al., 2009a), a particularly promising approach to stimulating professional conversations about teaching approaches was the framework illustrated in Figure 5.2.

In the teacher-demonstration approach, the teacher engages students in discussing an on-screen geometric construction and may ask questions about the objects on the screen to get the learners to explain what they might expect would happen if some parts of the configuration were moved or changed. This approach was found to allow teachers with little experience of using technology in the classroom to experiment with the technology with relatively small risk of things going wrong. In addition, this kind of use requires less change in the classroom setting and needs fewer resources than either organizing classes into a computer room or using a class set of laptops in the regular classroom. The second approach (in Figure 5.2) entails teachers providing previously created interactive files for their learners. With such teacher-created files, students can experiment with dynamic objects. This provides clear boundaries for learners and time is not spent setting up the tasks; rather, learners can spend time exploring the mathematics that is central to each task. No doubt there is quite some teacher control over the material, but the approach can bring in opportunities for creative thinking and problem solving by learners. The third approach (in Figure 5.2) involves learners creating their own files, perhaps

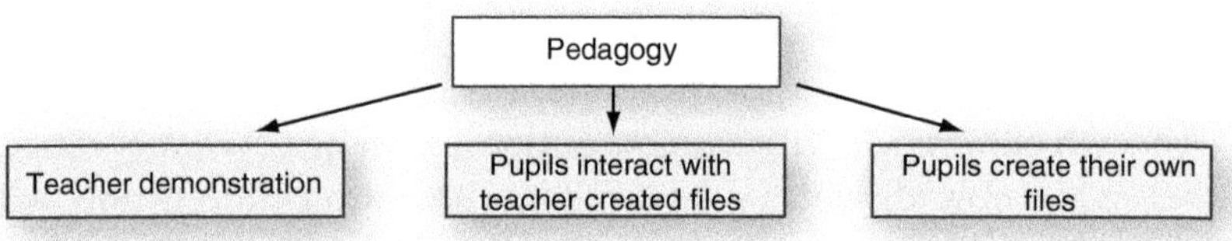


**FIGURE 5.2** Framework of Teaching Approaches with Geometry Software

for other learners to tackle. This approach provides some learner ownership of the work and engages a different sense of problem solving and thinking by creating that ownership. There is also the development of independence – in learning how to use the software, and with additional scope for student creativity and discovery. In the research project, this framework of teaching approaches provided both a way of structuring discussion during teacher CPD sessions, and a prompt for further discussion and further work – supported online through the provision of a wiki. All this points to ways in which teaching approaches developed by teachers can be used to stimulate further professional development by creating a network of teachers who are looking to enhance their sue of ICT in the teaching and learning of geometry.

Given these findings from research, it is instructive to note that the first issue of BJET, in 1970, included a review of Oettinger's (1969) book on the 'mythology of educational innovation'. In the book, Oettinger concluded that, at the time, 'education technology has not reformed – much less revolutionized – education' (ibid., p. 215). This prompted the BJET reviewer to observe that the major contribution of the book was that of 'alerting educators . . . to unrealistic technological expectations and heading off widespread disenchantment which might hinder the progress of educational technology' (Seatter, 1970, p. 79).

## Conclusion

The main message of this chapter is that the undoubted, and so far unparalleled, affordances of ICT, must be measured against the complexity of classroom learning, the demanding role of teacher and the need for relevant professional development. While access to computing power may shortly no longer be a source of an unbridgeable 'digital divide', differential access to networks of people that provide support for, and nurturing of, educational innovation via ICT may be an emerging new form of 'digital divide'.

As Oettinger warned some 40 years ago, it can be that expectations of technology appear unrealistic. Even with a classroom with suitable equipment, and with supportive institutional and national policies, such things may not be enough to counter the Ofsted claim about school teaching in mathematics in England makes a 'relatively limited contribution to developing pupils' ICT skills' (p. 27) and that 'despite technological advances, the potential of ICT to enhance the learning of mathematics is too rarely realised' (Oettinger, 1969). In developmental psychology the notion of canalization has been widely invoked, mostly in a nurture/nature argument (see e.g., Gottlieb, 1991). Yet the idea (captured by the term 'canalization') that there is a 'normal' pathway of development and that this can withstand 'great assaults or perturbations and still return to (or remain on) its usual developmental pathway' (ibid., p. 4) might have the potential to illuminate, at least to some extent, the issue at the heart of this chapter – that of why, despite the widely-acknowledged potential of ICT, integration into mathematics teaching and learning has proceeded much more slowly than some have predicted.

## References

Brown, M., Jones, K. and Taylor, R. (2003). *Developing Geometrical Reasoning in the Secondary School: Outcomes of Trialling Teaching Activities in Classrooms, a Report to the QCA*. London: QCA.

Christou, C., Jones, K., Mousoulides, N. and Pittalis, M. (2006). 'Developing the 3DMath dynamic geometry software: Theoretical perspectives on design'. *International Journal of Technology in Mathematics Education*, 13 (4), 168–74.

Christou, C., Sendova, E., Matos, J-F., Jones, K., Zachariades, T., Pitta-Pantazi, D., Mousoulides, N., Pittalis, M., Boytchev, P., Mesquita, M., Chehlarova, T. and Lozanov, C. (2007). *Stereometry Activities with DALEST*. Nicosia, Cyprus: University of Cyprus.

Gottlieb, G. (1991). 'Experiential canalization of behavioral development: Theory'. *Developmental Psychology*, 27, 4–13.

Hoyles, C and Jones, K. (1998). 'Proof in Dynamic Geometry Contexts'. In: C. Mammana and V. Villani (eds), *Perspectives on the Teaching of Geometry for the 21st Century* (ICMI Study 8), pp. 121–28. Dordrecht, The Netherlands: Kluwer Academic Publishers.

Jones, K. (2000). 'Providing a foundation for deductive reasoning: Students' interpretation when using dynamic geometry software and their evolving mathematical explanations'. *Educational Studies in Mathematics*, 44 (1–3), 5585.

—(2002a). 'Issues in the teaching and learning of geometry'. In Linda Haggarty (ed.) *Aspects of Teaching Secondary Mathematics*, pp. 121–39. London: Routledge.

—(2002b). 'Research on the use of dynamic geometry software'. *MicroMath*, 18 (3), 18–20.

Jones, K., Lavicza, Z., Hohenwarter, M., Lu, A., Dawes, M., Parish, A. and Borcherds, M. (2009a). 'Establishing a professional development network to support teachers using dynamic mathematics software GeoGebra'. *Proceedings of the British Society for Research into Learning Mathematics*, 29 (1), 97–102.

Jones, K., Mackrell, K. and Stevenson, I. (2009b). 'Designing digital technologies and learning activities for different geometries'. In Celia Hoyles and Jean-baptiste (eds), *Mathematics Education and Technology: Rethinking the Terrain* (ICMI Study 17), chapter 4, 47–60. New York: Springer.

Oettinger, Anthony G. (1969). *Run, Computer, Run: The Mythology of Educational Innovation.* Cambridge, MA: Harvard University Press.

Seatter, 1970. 'An essay of mythology'. *British Journal of Educational Technology*, 1 (1), 74–79.

Zachariades, T., Jones, K., Giannakoulias, E., Biza, I., Diacoumopoulos, D. and Souyoul, A. (2007). *Teaching Calculus Using Dynamic Geometric Tools.* Athens, Greece: University of Athens.

Chapter 6

# Learning Mathematics Using Digital Resources: Impacts on Learning and Teaching for 11- to 14-year-old Pupils

Don Passey
*Department of Educational Research, Lancaster University, UK*

## Introduction

This chapter explores impacts on the teaching and learning of mathematics when particular digital resources are used through the medium of interactive whiteboards. The impact of technologies on subject attainment and achievement (in mathematics as well as in other subject areas) has been a focus of a number of research studies, which have adopted a range of research approaches, both qualitative and quantitative. Enhanced subject attainment was identified in some studies (in Becta, 2001, 2003; Harrison et al., 2002), while Cox et al. (2003) reviewed reasons for these gains. When technologies are used in teaching and learning, a commonly reported gain by both teachers and pupils is enhanced motivation in subject activities. Some studies have identified aspects of specific positive motivational impacts (see Passey et al., 2004; Cox, 1997; Denning, 1997), and some instances studied have involved uses of digital resources. This chapter reports outcomes from a national evaluation studying motivational, learning and teaching gains when digital resources were used in mathematics lessons with 11- to 14-year-old pupils in a range of schools. The argument presented in this chapter is that digital resources provide enhancements for learners that are recognized as being motivating, that increased motivation can be linked to enhanced beliefs in abilities to undertake and engage with mathematics (in which discussion, enhanced understanding and conceptualization play important roles), and that specific ways in which learning is enhanced can be described by observers, teachers and pupils themselves.

## The Project Studied

Digital resources are often made accessible to pupils in classrooms through the medium of an interactive whiteboard. A 3-year project (the *Maestro Project*) was set up by Research Machines Education plc in 2003, initially involving 27 schools, to explore how digital resources (*RM MathsAlive* and mathematical resources available from other companies such as *MyMaths*), when used in a number of mathematics classrooms and departments, and supported externally, might impact on mathematics learning and teaching. The digital resources covered a complete curriculum for mathematics across Key Stage 3 (for pupils 11–14 years of age). Although resources were initially designed to support learners gaining average attainment, teachers accessed or developed resources to suit widely different pupil groups in terms of their mathematical knowledge and attainment. The digital resources, originally developed to meet the objectives of the *Secondary National Strategy framework for teaching mathematics* (DfEE, 1999), provided teachers with video openers (for watching and listening to), mental starters (tackling short problems with quick-fire or timed responses), specifically created interactive whiteboard screens (covering a specific topic or mathematical problem using features such as cover and uncover), main activities (up to about an hour in length and using additional physical resources such as counters or bricks), worksheets (mainly of a textual nature, for printing off and completion), games (using full multimedia and involving groups or teams competing against the clock or each other), and assessment exercises (designed to identify attainment levels). The resources were designed for easy access and use on interactive whiteboards, as well as for use in computer suites or computer clusters where learners could access resources more individually. Some graphical calculator activities were also available, and schools could set up pupil access beyond the school. Access to all resources was supported through an ICT-based management system, a virtual teaching and learning environment through which teachers could select digital resources.

## Background to the Study

Although research shows that features of digital resources can have a major impact on learning when interactive whiteboards are used, a review of the academic literature (Higgins et al., 2007), pointed out that explanations of how impacts were arising was sparse. While the distinction drawn by

Kennewell and Beauchamp (2007) between two sources of affordances ('those intrinsic to digital media and devices and those constructed by hardware designers, software developers and teachers preparing resources for learning') was difficult to make when identifying outcomes in this study, it was clear that specific digital resources offered specific affordances that impacted upon learning engagement and outcomes (argued further in Passey, 2006). While it can be argued that evidence about uses of a specific range of digital resources potentially limits the gathering of more circumspect findings, it can also be argued that specific digital resources need to be studied in depth in order to identify and understand in detail the specific learning impacts arising.

## Methodological Approaches

Evidence about uses and outcomes of digital resources was gathered through teacher interviews (from 19 key teachers and 19 heads of mathematics or senior managers), pupil interviews (open questions were asked of 57 pupils, 30 boys and 27 girls, largely in Year 9, across 16 schools), pupil questionnaires (with 20 questions posed in positive and negative ways to check reliability of responses, with 426 pupils providing responses, 164 girls and 251 boys, while 11 pupils declined to indicate gender), and lesson observations (in 18 lessons in different schools). Questions asked of pupils and teachers focused on: levels of use of resources; whether a curriculum scheme was used; how lessons were planned; perceived benefits and disadvantages of the resources; enjoyment; perceived impacts, whether expectations had been affected; impact on workload; perceived impact on connections to concepts, enhancements to understanding, retention or recall; perceived impacts on motivation, specific resources of value, impacts on specific individuals or groups, home use, application across year groups and the key stage and forms of support used and valued. Observations in classrooms focused on complementary issues: group features (size, gender mix and prior attainment); lesson objectives; classroom layout (the position of the interactive whiteboard, laptop or computer equipment, static whiteboards, windows and lights); timings and lengths of specific activities undertaken (the form of activity, forms of Information and Communications Technology (ICT) used, and forms of teacher and pupil interactions involved); evidence of metacognitive skills focused on or developed; evidence of enhanced discussion about mathematical strategies and evidence of transfer of learning (the providing of hooks or making links to prior or future learning).

## Impacts Reported by Learners

Overall, questionnaire responses from pupils were positive, with key responses indicating positive attitudes towards mathematics and the use of interactive technologies to support the learning of mathematics (largely supported both by teacher responses and independent observations). Pupil responses indicated that many had continued to enjoy mathematics, and that their enjoyment had increased over the period of the 3-year project; 63 per cent indicated that their enjoyment of mathematics had increased since the beginning of Year 7, 65 per cent indicated that their enjoyment had increased since the previous year, and 81 per cent indicated that they felt they could learn new things more easily than they could at the start of Year 7. These responses suggested that activities in these classrooms were facilitating enjoyment and positive forms of motivation (goal performance and academic efficacy). When asked whether they enjoyed using interactive whiteboards in mathematics lessons, 87 per cent of pupils agreed or agreed strongly. One link between levels of responses regarding enjoyment, positive motivations for learning, and enjoyment of uses of interactive whiteboards was suggested by pupils' responses to a question asking whether they found it easier to remember things when they used the interactive whiteboard; in total, 71 per cent agreed or agreed strongly. Whether these levels could be explained by impacts of certain types of resources was explored in other questions. Pupils were asked whether they enjoyed playing mathematical games, and 58 per cent agreed or strongly agreed. When asked if the playing of mathematical games helped them to remember things, 69 per cent agreed or agreed strongly. Questionnaire responses indicated that digital resources were having positive impacts on many pupils, and their learning of mathematics.

## How Digital Resources Were Used in Lessons

Observations showed that teacher practices could have a dramatic effect upon impacts on learning. In some lessons observed, uses of digital resources matched ways that pupils described learning being enhanced. From lesson observations and pupil interviews (although it should be noted that points raised by pupils were gained from open rather than closed questions, so frequencies of responses given might not be as high as those identified using closed questions), five particular ways in which teachers used digital resources were seen to influence impact, pupil involvement and gain:

**Enhancing visual clarity**

The fact that a shape was shown accurately, rather than as a sketch, was potentially significant. Most people (including groups of teachers when asked) remember a square as a clear four-sided object with equal sides, subtended by a 90° angle at each corner. This form of memorized image allows the 'rules of being a square' to be related through the memorized image. If a square is depicted as a sketch, then the rules of being a square need to be remembered separately and applied to this object. Remembering the object as a clear square means that the rules can be 'seen' more readily. Some pupils explained that digital resources offered 'exact and clear' images (5 pupils), and that having an exact image meant that they could remember 'a real image'. Some pupils said that digital resources were easier to read than when handwriting was done on a static board (3 pupils), others commented on the importance of colour and brightness in drawings (4 pupils), while yet others explained that the layout of topic details was better (2 pupils). Some pupils said the large size of objects allowed them to see things easily (3 pupils), while some explained that animations and diagrams helped their understanding through visualization (11 pupils) and 'simplicity' (1 pupil).

**Clarifying a process**

Some pupils referred to digital resources as being a point of reference that everyone could see (4 pupils), that enabled more discussion on topics (3 pupils), and that teachers explained more how to solve problems (1 pupil). The discussion of strategies and approaches to problems meant that pupils had the chance to clarify processes involved. Pupils not only had the chance to consider how others were tackling process, but also the chance to verbalize process. It is arguable (see Vygotsky, 1978) that verbalization is as important as (if not more important than), undertaking a series of examples in silence, written into exercise books.

**Developing conceptual understanding**

Some pupils explained that digital resources helped them understand certain topics, particularly shape (10 pupils), trigonometry (8 pupils), mental mathematics with addition, subtraction, multiplication and division (7 pupils), algebra (3 pupils), graphs (3 pupils), fractions and decimals (3 pupils), times tables (3 pupils), co-ordinates and grids (2 pupils), averages

(1 pupil), stem and leaf diagrams (1 pupil), and scatter diagrams (1 pupil). Other pupils explained that work was broken down into steps rather than being 'in big chunks in a text book' (4 pupils). When presented with 'loads of writing' this was found to be daunting, and pupils then believed that they could not do the work. Moving imagery was a key means for pupils to 'see what is meant', rather than their trying to 'imagine what is meant' when only described by teachers. Imagining what is meant can mean that pupils have to imagine the steps in a process, or the flow of a phenomenon. Using still and moving imagery meant that pupils could see these aspects, so understanding could be modelled rather than assumed.

**Encouraging participative learning**

Some pupils felt that games were interesting, and that this aspect alone helped them (12 pupils), while others felt it was 'good for exercising the brain' and 'drawing you into thinking' (2 pupils). Some pupils said that games activities helped all the class to work together, so that everyone was involved (6 pupils), so that pupils helped each other (2 pupils), and that competitive activity 'gets brains working' (3 pupils). Many pupils referred to different aspects of participative learning, whether this was in terms of increased opportunities for group discussion or all individuals from across a class taking part in games. Participative learning was important for two groups of pupils: those who had difficulty in participating because of low engagement levels and those who felt uncomfortable when some pupils were not engaged in lesson activities.

**Increasing pace and variety**

Some pupils said that they remembered things when they actively used digital resources but did not have to copy and write (17 pupils); they explained that they listened more (1 pupil), that it attracted them like a television would (1 pupil), and that it offered more variety (2 pupils). Pupils here were challenging the belief that copying and writing is the most vital means to support learning, when other alternatives exist. Some pupils felt the dynamics of lessons had changed, that they were conducted at a quicker pace, since the teacher had resources prepared and loaded in advance (7 pupils). Some pupils felt that a slow pace meant they lost focus and did not maintain attention. With a faster pace, they felt they could understand the flow better, and could easily go back to things they were not certain about (although there is need for some caution, as some teachers felt that some pupils did not benefit in this respect, especially those who needed more time to consolidate learning, and indeed other studies have pointed to pupils reporting increased pace to be a disadvantage (see Wall et al., 2005), while Smith et al. (2006) observed in their study that increased pace arose due to a higher level of questioning but with associated answers of a briefer nature). Pace enabled a better flow to be maintained, both for those who were distracted by breaks in flow, and for those where understanding was interrupted by breaks in flow. Using digital resources meant that different activities could be introduced more seamlessly, without long periods of waiting between one activity ending and the next one starting.

## Analysing the Width of Reported and Observed Impacts on Learning

Although digital resources impact on learning, it is clear that they do not impact to the same extent on each aspect of learning. A range of learning elements in framework format can be used to explore width of impact (shown in Figure 6.1). For this analysis, aspects of learning are grouped and detailed in six broad categories: megacognitive (using the work of Vygotsky, 1978; Bransford et al., 2000); cognitive (using the work of Bloom, 1956; Child, 1973; Gardner, 1991; DfES, 2006); metacognition (using the work of Presseisen, 2001); motivation (using the work of Passey et al., 2004); social interactions (using the work of Lave and Wenger, 1991; Twining and McCormick 1999); and societal implications (using the work of Lipman, 1995; Moseley et al., 2005). These aspects and categories are fully defined and described elsewhere (in Passey, 2008). Using evidence from pupils, teachers and observations, aspects of learning where impacts have been found are shaded in the framework.

From Figure 6.1 it is clear that certain aspects of learning are affected to greater extents than are others. Digital resources are impacting on:

### Megacognitive elements

Digital resources were not used generally to provide a 'big picture' for pupils. In terms of working within Zones of Proximal Development, many teachers selected resources and activities appropriate to the mathematical skills and understanding of pupils, or adapted resources to accommodate the needs of gifted and talented pupils or those with specific educational needs. While some teachers were concerned with transfer of previous learning into lessons, only a minority focused on transfers of learning to future situations beyond lessons.

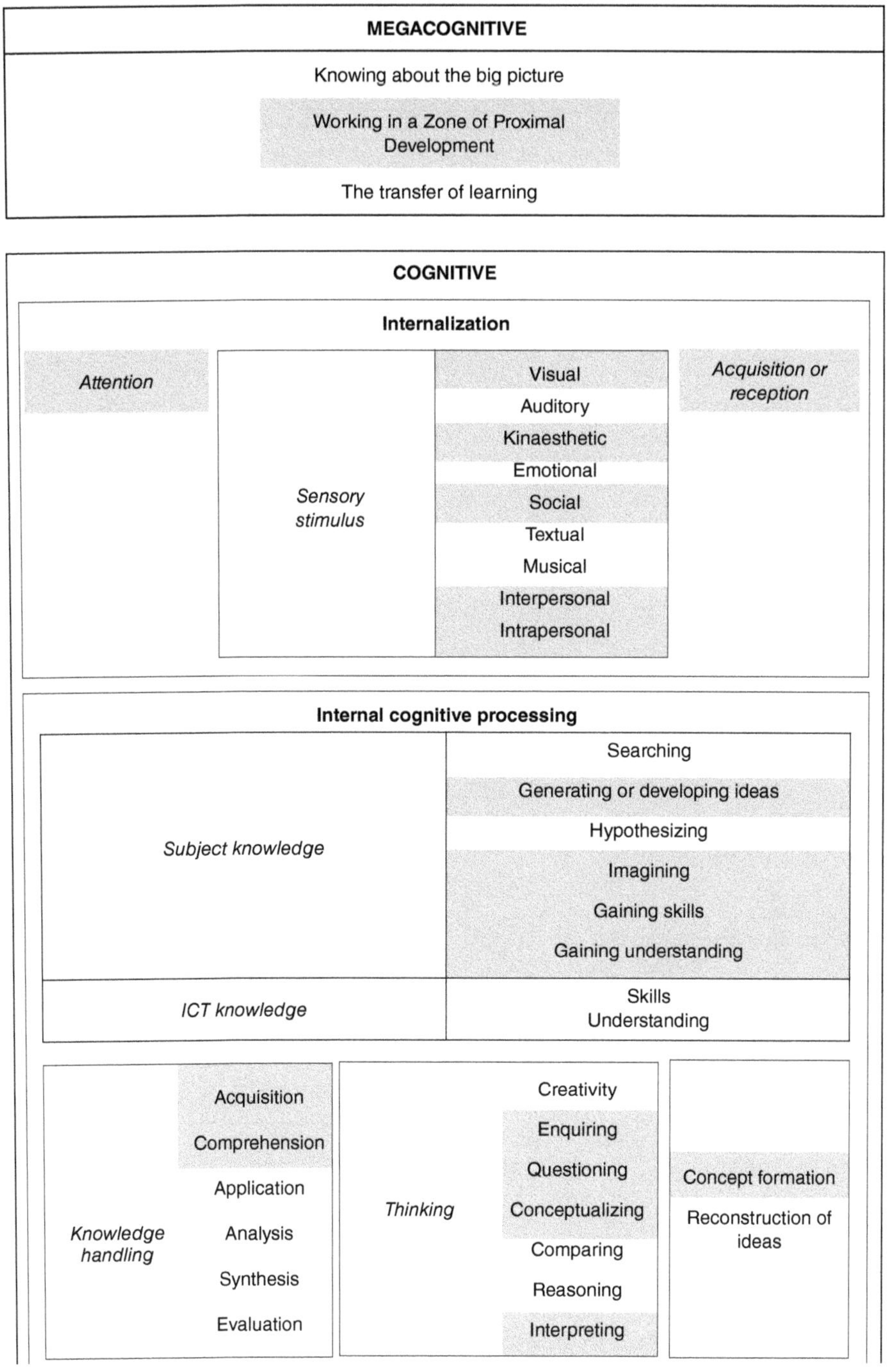


**FIGURE 6.1** Continued

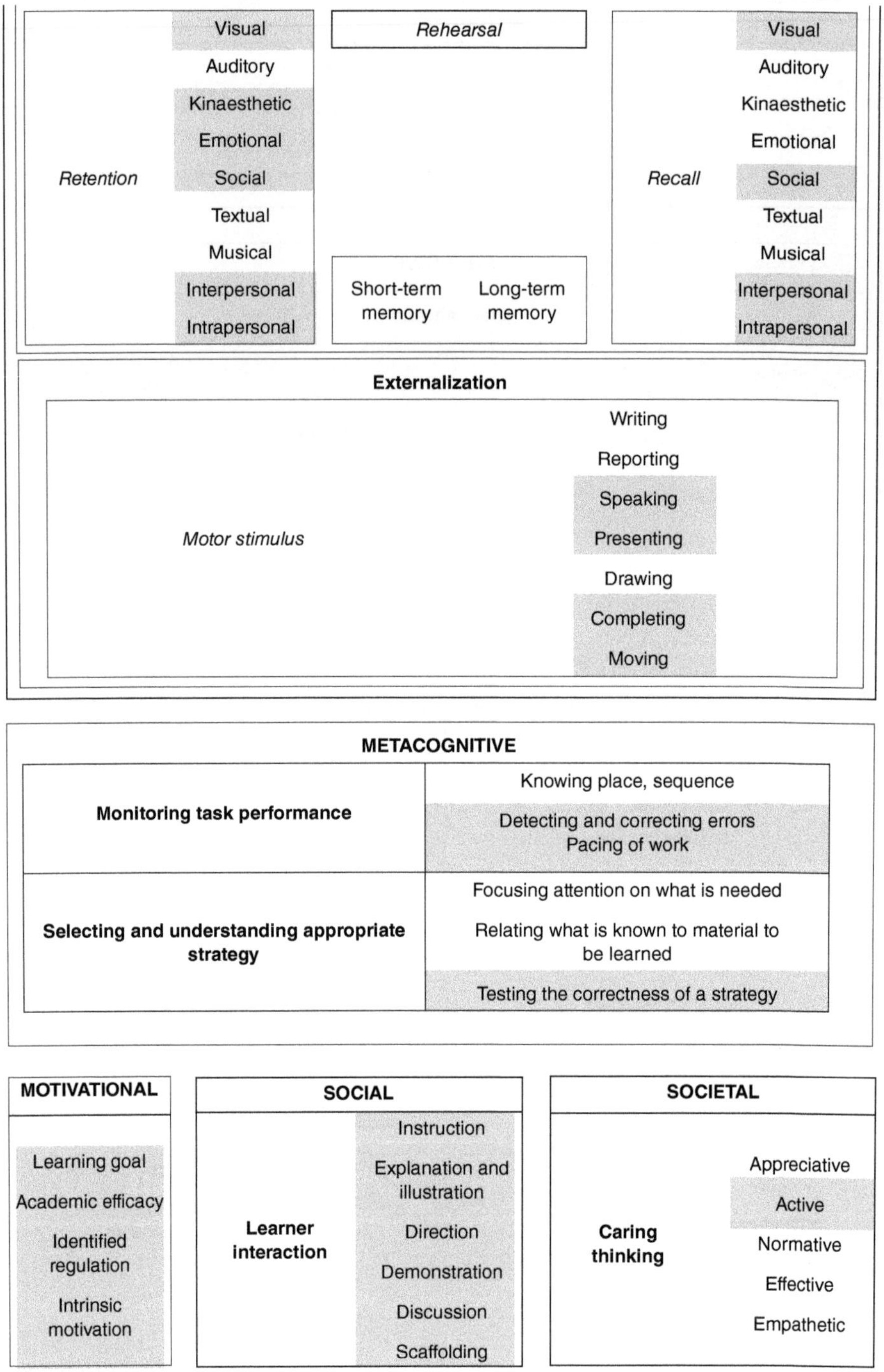


**Figure 6.1** Continued

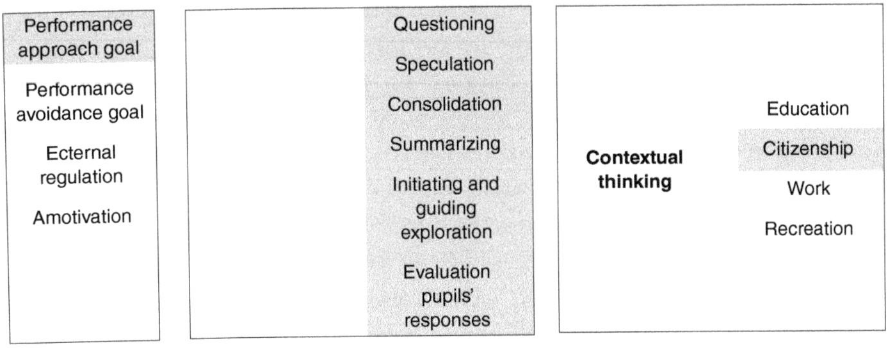


**FIGURE 6.1** Framework Showing Impacts of Digital Resources Used to Support Learning and Teaching of Mathematics

## Cognitive elements

**Multi-sensory learning elements:** Digital resources helped to maintain attention, a number of sensory routes were stimulated, and acquisition and reception of ideas or knowledge were encouraged. Attention was maintained because of size of images and text, clarity and detail, colour and movement. Handwriting by teachers alone on screens was much less effective, but handwritten annotations on prepared resources could readily highlight aspects for pupils. Uses of timers to count down, of sound to indicate when things had been achieved, of covered items that could be uncovered, and pupils touching the interactive whiteboard with pens or fingers to show how they had tackled problems, all added to the richness and diversity of the multi-sensory dimension.

**Knowledge acquisition:** The digital resources supported knowledge handling through clarity and visibility. Pupils recognized both the value of having clear images, and the importance of accessible examples at the time they were asked to think about a mathematical issue. Pupils could see images clearly when working with aspects of shape and space.

**Concept formation:** Clarity, visibility and visualization all helped with concept formation. Understanding what happened to the shape of a graph for the equation '$y = mx + 10$', when m was 0, 1, 2 or any other number, could be visualized more easily if the graph was very quickly generated and able to be compared to previous graphs. Simulations, animations and the comparison of results all supported the formation of concepts, explored also in Murphy et al. (1995), while Wall et al. (2005) report on visual effects helping learning.

**Higher order thinking skills:** In some lessons observed, teachers used digital resources in novel ways to explore an analysis of data. Digital resources allowed batches of data to be built up, with pupils directly involved in 'handling' the data, seeing what happened to the mean, mode and median when greater quantities were built up. Using resources, teachers engaged pupils in open discussion, and encouraged them to draw ideas from their existing knowledge, as well as asking them to indicate how far they accepted their peers' ideas and strategies for addressing problems.

**Knowledge retention and recall:** Clarity, visibility, and visualization leading to effective modelling and holding of rules and ideas appeared to have impact on knowledge retention and recall. Kozma (1994) refers to increased recall arising when imagery and diagrams are used alongside text: 'such structures are more memorable than those constructed with text'.

**Externalization:** Teachers moved away from using writing as the only (or major) form of externalization. Some teachers reported impacts on teaching, moving from pupil writing to more pupil discussion. Discussion and presentation were more prominently used by a number of teachers to offer effective ways to support externalization of knowledge and understanding. In some lessons observed, pupils came to the interactive whiteboard to show how they had undertaken a task, and to present this to the rest of the class.

### Metacognitive elements

In a few lessons observed, digital resources were used to offer pupils opportunities to find out if they were right or wrong, to address misconceptions and to identify how to do better next time. Teachers in some lessons were able to annotate screens, and to point out specific detail to help to focus the attention of pupils. The technology enabled pupils' work to be shared more easily; allowed them to show how they produced an answer and gave the opportunity to create and test general rules.

### Motivational aspects

Some pupils explained that work was broken down into steps, which was more motivating than being 'in big chunks in a text book'. Some teachers believed that pupils were motivated because they saw mathematical ideas presented through a 'recognized resource' as being akin to a 'universal truth'. Many pupils felt that they could tackle mathematics more readily when using digital resources, and some showed that their expectations could be raised. One pupil pointed out to another who was trying to create an angle of 90° using *Geometer's Sketchpad* that making an angle of 90.9° was 'poor'. The pace of lessons was also felt by pupils to help their motivation.

**Social interactions**

Some teachers believed digital resources allowed lines of questioning to be explored more. Haldane (2007) suggests that:

> While the recall of content from the computer's memory to the screen is not quite as rapid as the process of verbally articulating some knowledge held within the human mind, it is nevertheless fast enough to enable a striking intervention in the dialogue to be made in a way that commands attention and prompts further dialogue.

Pupils in some lessons were given opportunities to discuss questions or concepts in pairs or small groups. Teachers believed resources supported them in evaluating pupil responses; a teacher in one lesson asked pupils for their solutions first, and then only after exploring all pupil solutions was the teacher's own solution offered. Wide participation in lessons was clearly an important outcome for some pupils, particularly when games were used. As Haldane (2007) commented: 'Each pupil was moving the word with Martha, albeit cognitively and not physically'.

**Societal implications**

Certain societal aspects were introduced; teachers reported that having more application to real life situations had had impact on their teaching, and that video openers provided practical contexts for mathematics, which was important for pupils if they were to see value in mathematics that was wider and longer term, rather than just having immediate subject interest. Additionally, a range of pupils recognized the value of activities that were seen to 'care for others'.

## Conclusion

Many pupils reported that their enjoyment of mathematics had increased since the beginning of Year 7, that they could learn new things in mathematics more easily than at the start of Year 7, that they enjoyed using digital resources in mathematics lessons, that it was easier to remember things when they used them, and that playing mathematical games helped them to remember things (although slightly fewer indicated that they enjoyed playing mathematical games). Teachers used digital resources in five ways that were frequently reported by pupils as enhancing their engagement with and understanding of mathematics: to enhance visual clarity; to clarify a process; to develop conceptual understanding; to encourage participative learning and to increase pace and variety. The extent of impact of these five elements was not quantified by this study, and subsequent research could more accurately identify more precise levels of impacts.

Some teachers reported that the focus of their lessons had changed as a result of using digital resources; the focus had moved away from them as teachers towards more of a focus on the resources they were using. Some teachers felt digital resources enabled a more collaborative environment to emerge, where they were working with pupils to construct problems and devise methods to solve them. These forms of focus were moving teachers away, as one teacher said, from 'what to teach' to thinking about 'how to teach'. Some teachers reported that digital resources extended their teaching styles to incorporate more high-level discussion, group work and linking different areas of the curriculum within real-life contexts.

Pupils and teachers reported learning and teaching benefits when using digital resources, although some elements and aspects of learning were supported far more than others. The learning framework analysis identified a range of learning aspects that were not supported through uses of digital resources in this project. In terms of impact on attainment tests, some pupils believed that it helped them with Standard Attainment Tasks (4 pupils), and these pupils reported these responses prior to their results being known. Although this study shows that digital resources can support learning at a wide level, the proof that these features then lead to learning that can be assessed through examination procedures is yet to be shown.

## Acknowledgements

The author would like to thank Research Machines Education plc and the *Maestro Project* for support of this independent research. Further details and a summary report for teachers can be found at:

www.rm.com/Secondary/InTheNews/Article.asp?cref=MNEWS830586
www.rm.com/_RMVirtual/Media/Downloads/Final_Maestro_Project_Paper.pdf

## References


Becta. (2001). *The Secondary School of the Future – A Preliminary Report to the DfEE.* Coventry: Becta.

—(2003). *Secondary Schools – ICT and Standards. An Analysis of National Data from Ofsted and QCA.* Coventry: Becta.

Bloom, B. (1956). *Taxonomy of Educational Objectives.* New York: Longmans.

Bransford, J. D., Brown, A. L. and Cocking, R. R. (eds) (2000). *How People Learn: Brain, Mind, Experience, and School.* Washington, DC: National Academy Press.

Child, D. (1973). *Psychology and the Teacher.* London: Holt, Rinehart and Winston.

Cox, M., Abbott, C., Webb, M., Blakeley, B., Beauchamp, T. and Rhodes, V. (2003). *ICT and Attainment: A Review of the Research Literature. ICT in Schools Research and Evaluation Series No. 17.* Coventry/London: Becta/DfES.

Cox, M. J. (1997). *The Effects of Information Technology on Students' Motivation.* London: Kings' College London and Coventry: NCET.

Denning, T. (1997). *IT and Pupil Motivation: A Collaborative Study of Staff and Pupil Attitudes and Experiences.* Stoke-on-Trent: Keele University and Coventry: NCET.

Department for Education and Employment. (DfEE) (1999). *The National Numeracy Strategy: Framework for Teaching Mathematics from Reception to Year 6.* London: DfEE.

Department for Education and Skills (DfES) (2006). 'Thinking Skills web-pages, on the National Curriculum pages'. www.nc.uk.net/LACcs_thinkskill.htlm). (accessed on 6 November 2009)

Gardner, H. (1991). *The Unschooled Mind: How Children Think, How Schools Should Teach.* New York: Basic Books.

Haldane, M. (2007). 'Interactivity and the digital whiteboard: Weaving the fabric of learning'. *Learning, Media and Technology,* 32 (3), 257–70.

Harrison C., Comber C., Fisher T., Haw K., Lewin C., Lunzer E., McFarlane A., Mavers D., Scrimshaw P., Somekh B., Watling R. (2002). 'ImpaCT2: The impact of information and communication technologies on pupil learning and attainment'. A report to the DfES. ICT in Schools Research and Evaluation Series No.7. Coventry: Becta. www.becta.org.uk/research (accessed on 6 November 2009).

Higgins, S., Beauchamp, G. and Miller, D. (2007). 'Reviewing the literature on interactive whiteboards'. *Learning, Media and Technology,* 32 (3), 213–25.

Kennewell, S. and Beauchamp, G. (2007). 'The features of interactive whiteboards and their influence on learning'. *Learning, Media and Technology,* 32 (3), 227–41.

Kozma, R.B. (1994). *The Influence of Media on Learning: The Debate Continues.* www.ala.org/ala/aasl/aaslpubsandjournals/slmrb/editorschoiceb/infopower/selectkozmahtml.htm, (accessed on 23 September 2007).

Lave, J. and Wenger, E. (1991). *Situated Learning: Legitimate Peripheral Participation.* New York: Cambridge University Press.

Lipman, M. (1995). 'Caring as Thinking'. *Inquiry: Critical Thinking across the Disciplines,* 15 (1), 1–13.

Moseley, D., Baumfield, V., Elliott, J., Higgins, S., Miller, J. and Newton, D. P. (2005). *Frameworks for Thinking: A Handbook for Teachers and Learning.* Cambridge: Cambridge University Press.

Murphy, J. F., Jain, N. L., Spooner, S. A., Hassan S.W., Shnasse J. L., Metcalfe E. S., Schoening, P. A, Cunnius, E. L., and Frisse, M. E. (1995). 'Use of an interactive electronic whiteboard to teach clinical cardiology decision analysis to medical students'. *Journal of the American College of Cardiology*, 25 (2), Supplement 1 238A.

Passey, D. (2006). 'Technology enhancing learning: Analysing uses of information and communication technologies by primary and secondary school pupils with learning frameworks'. *The Curriculum Journal*, 17 (2), 139–66.

—(2008). 'The Role of Mobile Technologies in Moving "Digital Natives" to "Learning Natives" '. In J. Traxler, B. Riordan and C. Dennett (eds). Proceedings of the mLearn2008 Conference 'The Bridge From Text To Context'. Wolverhampton: University of Wolverhampton School of Computing and Information Technology.

Passey, D. and Rogers, C. with Machell, J. and McHugh, G. (2004). *The Motivational Effect of ICT on Pupils: A Department for Education and Skills Research Project 4RP/2002/050–3*. Nottingham: DfES.

Presseisen, B.Z. (2001). 'Thinking skills: Meanings and models revisited'. In A. L. Costa (ed.) *Developing Minds: A Resource Book for Teaching Thinking*, third edition, pp. 47–53. Alexandria, VA: ASCD Publications.

Smith, F., Hardman, F. and Higgins, S. (2006). 'The impact of interactive whiteboards on teacher-pupil interaction in the national literacy and numeracy strategies'. *British Educational Research Journal*, 32 (3), 443–57.

Twining. P. and McCormick R (1999). *Learning Schools Programme: Developing Teachers' Information Communication Technology Competence in the Support of Learning*. Milton Keynes: Open University.

Vygotsky, L. S. (1978). *Mind in Society: The Development of the Higher Psychological Processes*. Cambridge, MA: Harvard University Press.

Wall, K., Higgins, S. and Smith, H. (2005). '"The visual helps me understand the complicated things": pupil views of teaching and learning with interactive whiteboards'. *British Journal of Educational Technology*, 36 (5), 851–67.

## Chapter 7

# Improving on Expectations: Preliminary Results from Using Network-supported Function-based Algebra

Walter Stroup, Lupita Carmona and Sarah M. Davis
*The University of Texas at Austin, US*

This chapter reports preliminary analyses comparing results on the state-administered 8th grade and 9th grade algebra Texas Assessment of Knowledge and Skills (TAKS) for a treatment and a control group. The treatment group consisted of 127 students from algebra classes at a highly diverse school in central Texas taught by two relatively new teachers using a network-supported function-based algebra (NFBA) approach as integrated with the ongoing use of an existing school-wide algebra curriculum. The control group was comprised of 99 students taught by two more-senior teachers in the same school using only the school-wide algebra curriculum. The intervention consisted of implementing 20–25 class days' worth of NFBA materials over an 11-week period in the spring of 2005. Because the students were not randomly assigned to the classes, the study is a quasi-experimental design. Using a two sample paired t-Test for means, statistically significant results for the treatment group (p-value one tail = $0.000335 > \alpha = 0.05$) were obtained. We can conclude the NFBA intervention was effective in improving outcomes related to learning the algebra objectives assessed on the 9th grade TAKS.

## Introduction

To date, the multiple-strands based approach to curricula promoted by the National Council Of Teachers of Mathematics (1989, 2000) has not displaced the single-strand Algebra I course as gatekeeper in the educational system of the United States. If anything, the standard, 'standalone', Algebra I course is now even more central at many levels, including in state curricula (e.g., minimum course requirements and exit exams) and in nationally administered tests (e.g., the new SAT tests). As a result, improving student outcomes related to the content of the traditional Algebra I curriculum is, perhaps, the single most strongly felt need relative to secondary mathematics education. Given the raised expectations regarding introductory algebra, we look to ask if there are ways of systematically improving on expected student outcomes in ways that move beyond the current overemphasis on addressing performance shortcomings with remediation. Our study looks to move in this direction. As illustrated by the results for our control group, past student performance on state-administered tests tends to be predictive of future testing outcomes. In our effort to identify approaches that are likely to improve expected student outcomes, not maintain them, we compared paired 8th and 9th grade TAKS results for the students in our study and asked the question: Do the students in our treatment group outperformed their peers in the control group on the algebra objectives tested on the state administered, 9th grade, TAKS? Did the NFBA intervention have the effect of improving on expected student outcomes?

Our intervention centred on the use of function-based algebra (FBA) as supported by generative activity design in a next-generation classroom network technology (i.e., the TI-Navigator 2.0 network combined with classroom sets of TI-84 Plus calculators). We call this approach network-supported function based algebra (NFBA). After providing some background for our study we report our results. Because the students were not randomly assigned, the study is based on a quasi-experimental design. Using a two sample paired t-Test for means, statistically significant results in outcomes for the treatment group (p-value one tail = $0.000335 > \alpha = 0.05$) were obtained.

## Background

There are three strands of analysis that are brought together in framing our study: (1) using FBA in a way that speaks more directly to the *structural* aspects of a standard introductory algebra curriculum, (2) situating this version of a function-based approach relative to *generative activity design* as supported by the capabilities of next-generation classroom networks (Stroup et al., 2005) and (3) explaining our use of performance on previous high stakes mathematics testing to evaluate the effectiveness of the algebra-specific interventions implemented for this study.

## Function-based Algebra Revisited – Emphasizing Mathematical Structure

In ways that highlight the idea of function, affordable technologies such as the graphing calculator have long been recognized to have the potential to substantively alter the organization of teaching and learning algebra concepts. Indeed, a number of approaches to pursuing FBA are discussed in the research literature (for an overview see Kaput, 1995). Many of these approaches were developed as part of an ambitious, and still ongoing, effort to fundamentally reorganize school-based mathematics to focus on *modelling*. For curricula, this would mean that the formal set-theoretical approaches to defining function that had come to be associated with the 'new math' movement would be downplayed and largely replaced by an approach highlighting how functions can be used to model co-variation – that is, how one variable is related to, or co-varies with, another variable. Computing technologies like the graphing calculator were to support significant engagement with, and movement between, *representations of functions* in symbolic, tabular and graphical forms. Indeed a technology-supported engagement with these 'multiple representations' of functional dependencies, especially as situated in motivating 'real world' contexts, has come to typify both *what function-based algebra is* and *why function-based algebra it is expected to be effective* with learners. In the United States this modelling-based approach to FBA informed the development of the 'standards-based' mathematics curricula funded by the National Science Foundation and then incorporated into various levels of 'systemic reform' initiative also supported by NSF. These systemic reform initiatives, anticipating the language associated with the more recent No Child Left Behind legislation, were to 'raise the bar' and 'close the gaps' in student performance. The significance of this modelling-focused alignment notwithstanding (e.g., the State Systemic Initiative in Texas played a considerable role in the State-wide adoption of graphing technologies for algebra instruction and assessment), in day-to-day practice a modelling-focused approach to FBA has fallen well short of displacing much of what still constitutes the core of traditional algebra instruction. In part, the feedback from educators seems to be that as powerful as 'real world applications' might be in motivating some students, the 'bottom line' is that abstractions and formalisms are what continue to be emphasized on standardized exams and thus are what teachers feel considerable pressure to engage. Among school-based educators who are feeling enormous pressure to improve testing outcomes, modelling-based FBA is simply not seen as sufficiently helpful in addressing the core 'structural' topics of a standard algebra curriculum. In framing our study, however, it is important to underscore that this perceived shortcoming is *not* a limitation in the potential power of using a function-based approach, but is only a limitation in a particular implementation of FBA that is, itself, principally motivated by the goal of making modelling the overarching focus of school-based mathematics (and far less by improving outcomes related to learning introductory algebra). For our study we take the strong position that while emphasizing modelling should continue to be important, a function-based approach also has enormous potential to improve student understanding of the structural aspects of introductory algebra. To make this point both with teachers and in our materials development, we have found it helpful to advance the following deliberately provocative, but still sincere, claim: *When viewed through the lens of a larger sense of what FBA can be, nearly 70 per cent of a standard introductory algebra curriculum centres on only three big topics.* These three big topics are: *equivalence* (of functions); *equals* (as one kind of comparison of functions) and a systematic engagement with aspects of *the linear function.* This approach, as it is to be investigated in this study, builds on ideas associated with FBA introduced by Schwartz and Yerushalmy (1992) (see also Kline, 1945).

A major strength of this more structurally-focused, FBA is that it allows for consistent interpretations of both *equivalence* and *equals* in ways that students can use to understand the seeming ambush of 'rules for simplifying' and 'rules for solving' typically presented early-on in a standard algebra curriculum. If the expression $x + x + 3$ is *equivalent* to the expression $2x + 3$, then the function $f(x) = x + x + 3$ and the (simplified) function $g(x) = 2x + 3$, when assigned to Y1 and Y2 on the calculator, will have graphs that are everywhere coincident. They will also have paired values in the tables that are, for any values in the domain, the same. Students will say 'the graphs' are 'on top of each other.' This 'everywhere the sameness' associated with equivalence then will be readily distinguished from *equals*, as just one kind of comparison of functions. Equals comes to be associated with the value(s) of the independent variable where the given functions intersect (and > is associated with where one function is 'above' another; < where one is 'below'). The students will understand from looking at the graphs that the function $f(x) = 2x$ and the function $g(x) = x + 3$ are clearly *not* equivalent (they are not everywhere the same). But there is one value of x where these functions will pair this x with the same y-value (the students will say there is one place where the functions are 'equal' or 'at the same value'). Graphically, equals is represented as the intersection in a way that is quite general and that readily extends beyond comparisons of linear functions (e.g., $-x2 + 2x + 8 = x2 - 4x + 4$). This distinction between equivalence and equals is helpful because in a standard, non-function based, algebra curriculum rules for simplifying expressions and rules for solving systems of equations are introduced very near each other and, not surprisingly, often become confounded. In addition students will feel like they have no ready way of checking their results, other than asking the teacher. In marked contrast, using a function-based approach, as supported by the use of a combined graphing, tabular and symbolic technology like a graphing calculator, students can readily 'see' the difference between these ideas and can use these insights to make sense of results from 'grouping like terms' as distinct from 'doing the same thing to both sides'. This then allows the students to test their own results, using the technology, for either simplifying or solving. For simplifying they can ask themselves if the resulting simplified function is everywhere 'the same' as the given function? For solving systems of linear equations they can ask did their attempts to 'do the same thing' to the linear functions on both sides of the equation preserve the solution set (i.e., the x-value at the intersection)? Having students be able to distinguish and make sense of these two core topics in a standard algebra curriculum is significant and illustrates the power of FBA to help with structural aspects of a standard Algebra I curriculum. These ideas were emphasized in the materials we developed. Of course, a modelling-oriented approach to FBA can be helpful in supporting student understanding of the third of the big three topics: a systematic engagement with aspects of the linear function. But herein we want to continue to illustrate some elements of a less modelling centric engagement with FBA. As a result, we will illustrate implementing aspects of studying linear functions using generative activity design as supported by new network technologies. The effectiveness of this structural approach to FBA, without network capabilities, has begun to be established (cf., Brawner, 2001). We now move on to consider the role new network technologies can have in further enhancing FBA.

## Supporting Generative Design with TI-Navigator TM 2.0

Briefly, generative design (cf. Stroup et al., 2005) centres on taking tasks that typically converge to one outcome, for example, 'simplify 2x + 3x', and turning them into tasks where students can create a space of responses, for example, 'create functions that are the same as f(x) = 5x'. The same 'content' is engaged for these two examples, but with generative design a 'space' of diverse ways for students to participate is opened up, and the teacher, based on the responses, can get a 'snapshot' of current student understanding (so, for example, if none of the functions the students create to be same as f(x) = 5x involve the use of negative terms, the teacher can see in real time that students may not be confident with negative terms and can use this information to adjust the direction of the class). To illustrate how generative design and NFBA can help with the third of the three core topics in a standard algebra curriculum, we'll briefly sketch some of the activities we used in our intervention. The Navigator TM 2.0 system allows students to move an individual point around on his/her calculator screen and also have the movement of this point, along with the points from all the other students, be projected in front of the class. In one introductory activity students are asked to 'move to a place on the calculator screen where the y-value is two times the x-value'. There are many places the students can move to in satisfying this rule, and this is what makes the task generative. Often the majority of the points are located in the first quadrant and this gives the teacher some sense of where the students are in terms of confidence with negative x- and y-values. This exploration of a rule for pairing points does describe a function and this approach to creating functions is not dependent on co-variation (indeed, should the teacher want to discuss it, this activity can be used to highlight a set-theoretic approach to defining a function). After observing that 'a line' forms in the upfront space, all the points then can be sent back to the students' calculators and can act as 'targets' for creating different functions on their calculators (in Y1= , Y2 =, etc.) that include ('go through') these points. Then the students can send up what they consider their 'most interesting' functions. A space of often quite interesting equivalent expressions is thereby created and shared in the upfront-space. To further explore ideas related to linear functions, students also can be given a rule like 'move to a place where your x-value plus your y-value add up to 2'. Again a 'line' forms but now when the points are sent back to calculators, the students are pushed to explore ideas related to moving from a linear function in standard form (i.e., x and y summing to 2) to the same function being expressed in slope-intercept form (the form the students must use on the calculators in order to send a function through the points). Again, these and many other structural ideas found in a standard algebra curriculum can be explored using NFBA.

## Improving on Expectations

As is mentioned earlier, the intent of the 'No Child Left Behind' legislation in the United States is to 'raise the bar' of what is expected of all students and to 'close the gaps' in performance of currently underserved populations. The effort is to be forward looking as higher expectations and measurable progress are to present a tight system of positive feedback in driving demonstrable improvement in educational outcome. Even in a time of heightened political partisanship in the United States, this vision is still seen as compelling and potentially unifying. But as systems theorists (cf., Senge, 1994) are quick to remind us, a challenge in implementing major structural reforms is ensuring that the intended dynamics meant to both characterize and drive the change – in this case positive forms of feedback between raised expectations and measurable outcomes – are not themselves overwhelmed by unanticipated and unintended consequences of what may be well-intending implementation. Relative to learning algebra, one widely used strategy is to preserve the current approaches to teaching algebra and then address shortcomings in student outcomes with remediation. The problem is that remediation, almost by definition, is an inherently *backward* looking and *corrective* strategy. Its role is to fix what is seen as broken, not to drive forward progress. Relative to mathematics education, with more and more effort at each grade level (especially in underperforming schools but also in lower 'tracks' in higher performing schools) spent on correcting for past or anticipated shortcomings (e.g., 'reviewing' material not mastered from previous years, funding remediation classes during the school year and/or in the summer, or spending considerable class time practicing test-taking skills) attention to proactive strategies (strategies that improve on expected outcomes) is being compromised. From a structural point of view a *positive* feedback loop – like that between raised expectations and measurable progress found at the heart of the NCLB legislation – needs practical *forward looking* and *forward acting* strategies to be effective. To make the case for NFBA being an example of one such strategy, we look to compare our treatment group outcomes on the 9th grade TAKS™ algebra objectives relative to what might be expected based on previous performances on the 8th grade TAKS.

## The Study

### Research Question

Does the network-supported function-based approach outlined above improve the performance of the treatment group in statistically significant ways relative to the performance of control group peers?

### The Sample

The study participants were 226 students from a diverse high school in central Texas. All the students were enrolled in 'non-repeater' (non-remedial) sections of Algebra I and nearly all the students were in 9th grade. Two relatively junior teachers were assigned by the department chair to the experimental group and two more-experienced teachers were assigned to the control group: 127 students were in the treatment group and 99 students were in the control group.

### Activities

In their Algebra I class, the treatment groups used a NFBA over 9 weeks of instruction in the spring of 2005. The treatment and control groups kept their curricula on the same topics but the experimental group used the NFBA materials, on average, approximately 2 days a week.

### Methods

The raw 8th and 9th grade scores for the State-administered TAKS tests were obtained for the students participating in the study. The 8th grade TAKS was taken before the intervention and the 9th grade TAKS scores for the algebra objectives were collected after the intervention.

### Analyses

The raw scores on the 8th grade TAKS and the algebra items on the 9th grade TAKS were converted to per cent correct results. Table 7.1 and Figure 7.1 show the comparison of the means for the 8th and 9th grade TAKS for the treatment and control groups.

**Table 7.1** Mean TAKS Score Results for Treatment and Control Groups

| | **Treatment** | **Control** |
|---|---|---|
| 8th GRADE TAKS SCORES | 53.8 | 56.4 |
| 9th GRADE TAKS SCORES (Algebra Items) | 57.9 | 56.1 |

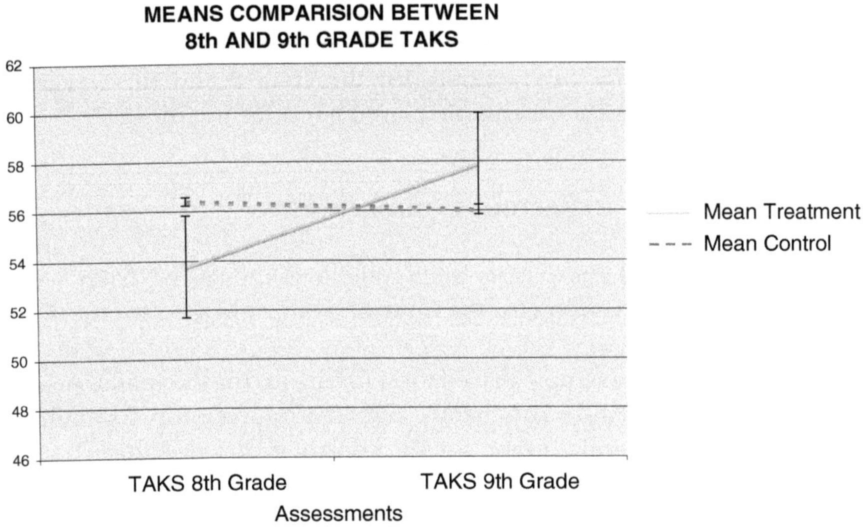


**FIGURE 7.1** Means Comparison of TAKS Results

We implemented *two* approaches to study changes attributable to the intervention: (1) comparing the student performances between the treatment and control groups first before the intervention (8th grade TAKS) and then after the intervention (9th grade TAKS for Algebra Items) (2) comparing the *paired* student performances before and after the intervention for the control group and then the treatment group.

## First Comparison

Our first approach was to carry out a comparison between Treatment and Control group results, first before and then after the intervention.

No statistical difference was found between treatment and control groups' results either before or after the intervention. The graph in Figure 7.1, however, suggests a need for additional analyses. On the graph it is clear that, although no statistically significant differences were found using the given methods, the treatment group started off about 2 per cent lower than the control group on the average 8th grade TAKS scores. Then after the intervention the plot of the 9th grade results shows that the students in the control group maintained almost the same average on the 9th grade TAKS score (the dotted line is almost completely horizontal, showing no change) whereas the treatment group's graph shows appreciable improvement, approximately 4 per cent. This suggests the possibility of comparing *paired* scores before and after intervention, for the control and the treatment groups separately, using a two-sample paired t-Test for the means.

## Second Comparison

Our second approach was to carry out a comparison of Paired TAKS Scores before and after intervention for the Control Group and then for the Treatment Group.

We performed a two sample paired t-test for means for the *control group* to look for changes in TAKS scores before and after intervention. As might be suspected from examining graph for the control group in Figure 7.1, the results of the t-test show no evidence that the means for the control group before and after the intervention are different (p-value one tail = $0.402 > \alpha = 0.05$). As a result we can conclude that the students in the control group maintained consistent averages for the 8th grade and 9th grade algebra TAKS scores. There was no statistically significant improvement. This result is consistent with the sense that absent changes in practice, performance in one year is likely to predictive of performance in the next. When we implemented a two sample paired t-test for the means for the treatment group, the results (p-value one tail = $0.000335 > \alpha = 0.05$) provided strong evidence of differences in means before and after intervention. This suggests the students in the treatment group improved significantly in paired results on the 8th and 9th grade TAKS. Considering that the treatment and control groups were comparable, that no improvement was shown for the paired 8th and 9th grade TAKS scores in the control group, and that improvement was shown for the paired 8th and 9th grade TAKS scores in the treatment group, we have strong evidence to say that this improvement in TAKS scores was an effect of the intervention. NFBA does appear to have been proactively effective in improving student outcomes.

## References

Brawner, B. F. (2001). 'A function-based approach to algebra: Its effects on the achievement and understanding of academically-disadvantaged students'. Unpublished doctoral dissertation, The University of Texas at Austin.

Kaput, J. J. (1995). 'A research base supporting long term algebra reform?' In D. T. Owens, M.K.Reed and G.M.Millsaps (eds), *Proceedings of the Seventeenth*

*Annual Meeting of the North American Chapter of the International Group for the Psychology of Mathematics Education*, vol. 1, pp. 71–94. Columbus, OH: ERIC Clearinghouse for Science, Mathematics, and Environmental Education. (ERIC Document Reproduction Service No. ED 389 539).

Klein, F. (1945). *Elementary Mathematics from an Advanced Stand-point: Arithmetic, Algebra, Analysis*. New York: Dover.

Schwartz, J. and Yerushalmy, M. (1992). 'Getting students to function in and with algebra'. In G. Harel and E. Dubinsky (eds), *The Concept of Function: Aspects of Epistemology and Pedagogy*. MAA Notes, 25, pp. 261–89. Washington, DC: Mathematical Association of America.

Senge, P. M. (1994). *The Fifth Discipline: The Art and Practice of the Learning Organization*. New York: Doubleday.

Stroup, W. M., Ares, N. and Hurford, A. (2005). 'A dialectic analysis of generativity: Issues of network supported design in mathematics and science'. *Journal of Mathematical Thinking and Learning*, 7 (3), 181–206.

Stroup, W. M., Carmona, L. and Davis, S. (2005). 'Improving on Expectations: Preliminary Results from Using Network-Supported Function-Based Algebra'. In S. Wilson (ed.), Proceedings of 27th Annual Meeting of the Psychology of Mathematics Education – North American Chapter, Blacksburg, VA.

TAKS is a trademark of the Texas Education Association.

TI-Navigator is a trademark of Texas Instruments Incorporated.

# Part Three

# Key Pedagogical Issues in Embedding ICT in Teaching and Learning Mathematics

One of the issues for teachers when using technology in the classroom is that of ensuring that the time invested in learning to use a digital tool, whether by the teacher or pupils, pays dividends in terms of enhancing learning. Moving away from the short-lived motivational benefits of using something new, in this chapter, authors reflect on the ways in which Information and Communications Technology (ICT) can truly contribute to learning. Colette Laborde considers the nature of tasks and problems for which dynamic geometry software is inherently useful and how learners can benefit from using it. John Mason writes about the use of manipulatives and how users can meaningfully interact with them. He contemplates the different levels of engagement and sophistication involved in experiencing and being convinced by phenomena, justifying them and using tools to recreate them. Rosemary Deaney and Sara Hennessy present their findings from a case study in which a teacher utilized a range of pedagogical strategies to capitalize on the opportunities afforded by different digital tools. It is interesting to consider how some of the strategies employed by the teacher in Deaney and Hennessy's study relate to the 'levels' of engagement with manipulatives discussed by Mason.

Chapter 8

# Designing Substantial Tasks to Utilize ICT in Mathematics Lessons

Colette Laborde
*University Joseph Fourier and IUFM of Grenoble, France*

## Introduction

The design of tasks is a complex activity involving many dimensions. Most of the time, mathematics teachers do not design tasks from scratch. They choose tasks from textbooks or from available resources and may adapt them to function, taking into consideration the variables in each classroom set up (tasks already carried out, students' available knowledge, notions already known . . . ). When tasks are technology based, the situation is even more complex, as adding technology deeply affects the task itself. A task planned for paper and pencil cannot simply be applied to that done with technology with unchanged learning aims. It is well known that some paper and pencil problematic tasks may become trivial when solved with technology, such as obtaining the graph of a function in a default window on a graphic calculator. Conversely technology offers affordances for new kinds of tasks not possible in paper and pencil environments specifically designed for fostering learning.

In a past research and development project (Laborde 2001) in which a group of teachers wrote teaching scenarios based on Cabri for high school, we could distinguish four kinds of tasks making use of Cabri and observe how tasks designed by teachers evolved over time.

## New Kinds of Tasks Made Possible by Technology

Four kinds of tasks among tasks designed by several teachers integrating Cabri in their teaching at high school were distinguished:

- tasks in which the environment facilitates the material actions but does not change the task for the students, for example, producing figures and measuring their elements;
- tasks in which the environment facilitates students' exploration and analysis, for example, identifying relations within a figure through dragging. A polygon and its translated image are given in Cabri. Students are asked to conjecture relations between the sides of the initial polygon and its image. In this kind of task, a visual recognition of the relations should be relatively easy. The role of Cabri is to help students make conjectures about the relations using the drag mode;
- tasks that have a paper and pencil counterpart but can be solved differently in the environment, for example a construction task may be solved in dynamic geometry environment by using a geometric transformation;
- tasks that cannot be posed without the mediation of the environment, for example, reconstructing a dynamic diagram through experimenting with it in order to identify its properties. A task like reconstructing a dynamic diagram given in Cabri (so called 'black box' tasks) takes its meaning from the Cabri environment itself, in particular from the drag mode which preserves geometrical relations. Such tasks require identifying geometrical properties as spatial invariants in the drag mode and possibly performing experiments with the tools of Cabri on the diagram.

Let us illustrate more in detail each kind of task with examples in the dynamic geometry environment Cabri for plane geometry (Cabri II Plus) or 3D geometry (Cabri 3D).

## An Example of an Unchanged Task: The Sum of the Distance of a Variable Point to Fixed Points

Students were asked to construct the figure displayed below (Figure 8.1), in which P is a variable point of segment AB. They were asked to measure PC and PD for several values of AP and then to study, in the paper and pencil environment as normal, the function PC+PD by making a table of values and drawing the graph. The teacher did not mention the possibility of obtaining this graph with Cabri as a locus which was a good reason to connect geometry with calculus.

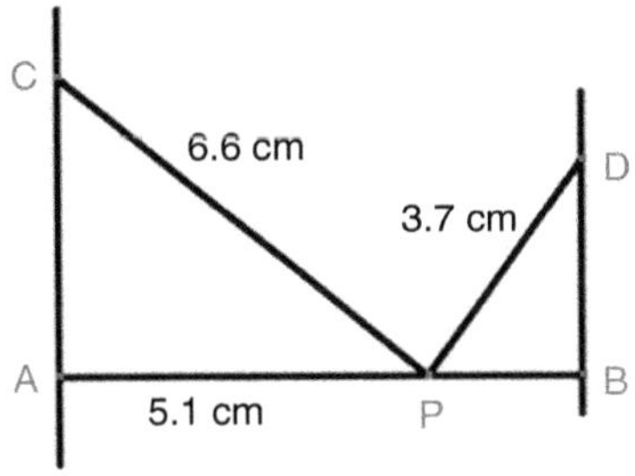


**FIGURE 8.1** Using Cabri for Easy Providing of Data

In this task, the software was mainly used as a provider of data. It was *neither the source of the tasks* given to the students, nor *part of the solving process* of these tasks that were intended to be solved without the computer.

## An Example of a Task with Possible Change in the Solving Process

In Cabri3D like environments, there are new strategies based on breaking down the cube into 2D elements. A lateral square can be considered as the image of the base in a rotation with axis the edge of the base plane (Figure 8.2). The other lateral faces can be obtained as images of the previous one in rotations around the vertical axis of the cube (Figure 8.3). Those strategies are not spontaneously used by students. We hypothesize that the construction strategies learned in 2D geometry become obstacles to new strategies specific of 3D.

We consider these strategies which are made possible by Cabri3D as interesting from a learning point of view, for two reasons:

- they enlarge the scope of possible ways of structuring the cube
- and they use construction tools based on objects and properties of 3D geometry and as such contribute to the learning of 3D geometry.

In a paper and pencil geometry, these objects and properties are not operational for construction tasks, they are only operational in proofs. The strength of Cabri3D like environments is that those objects and properties become operational construction tools. They can be used in action in construction tasks before being used at the level of proof. In construction

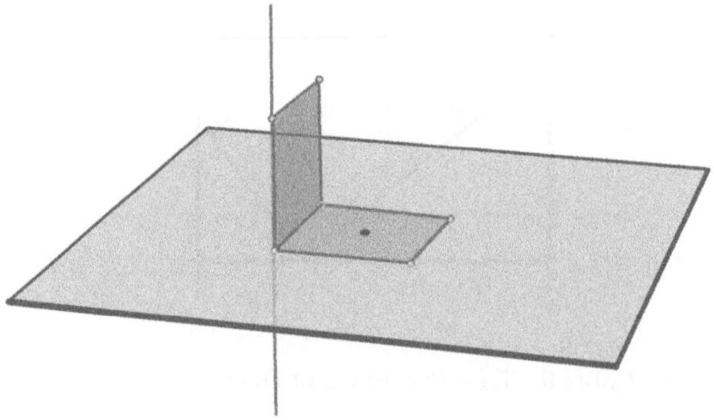

**FIGURE 8.2** A lateral face as rotated from the initial square

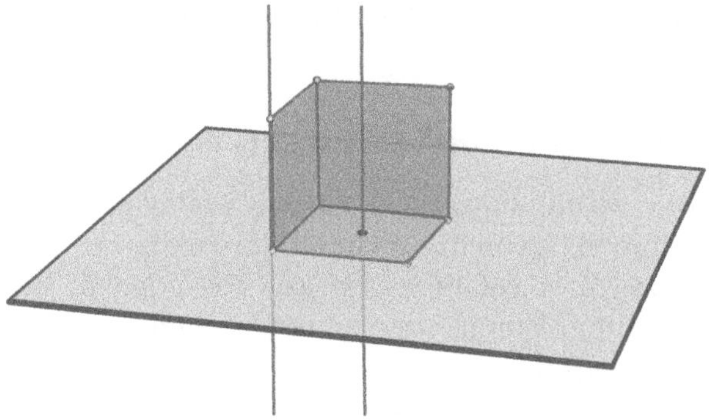

**FIGURE 8.3** The second lateral face as rotated from the previous one

tasks, students can observe that these strategies provide the expected result. The visual feedback strengthens the power of these properties.

Very relevant and beautiful examples of original and efficient instrumentation of Cabri3D are provided by Chuan (2006)[1]. They come from the lecture given by Chuan at ATCM[2] 2006 entitled 'Some unmotivated Cabri3D constructions'. 'Unmotivated' was explained by Chuan as 'non algebra, non routine, not found in Euclid, discovered accidentally, tailor made, so short, so beautiful, so fun'. These constructions are non-routine and not found in Euclid because the tools they required were not available. Chuan insists on the efficiency of the constructions ('short'). This is a critical feature of problems that are able to promote learning of new knowledge according to the theory of didactic situations (Brousseau

1997). A new solving strategy is likely to be constructed by an individual when his/her routine or available strategies are tedious or inoperative for the problem. The beauty of the solution emerges from the conjunction of its efficiency and its unusual character. Let us comment one of the examples given by Chuan: the triangular cupola starting from a hexagon (Figure 8.4).

Constructing the cupola requires analyzing it. The base of the cupola is a given regular hexagon and the top is an equilateral triangle. How to determine the distance of this triangle to the base plane? Actually the vertices of the triangle are vertices of regular tetrahedra (Figure 8.5). In this example, a regular tetrahedron is used as measurement transfer tool. This solution is based on a deconstruction of the cupola into 3D and 2D components of the figure.

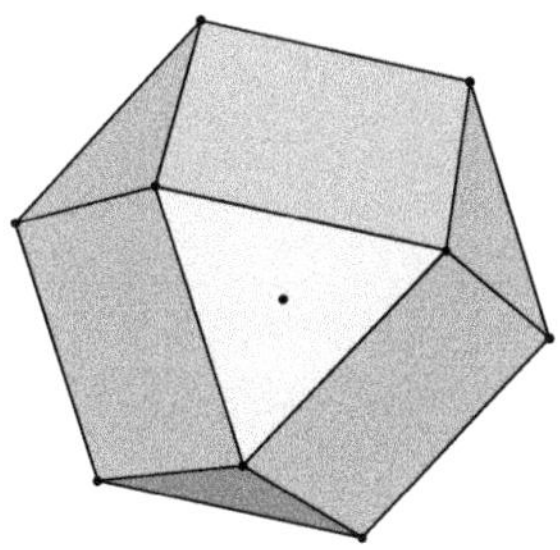

**FIGURE 8.4** A Triangular Cupola

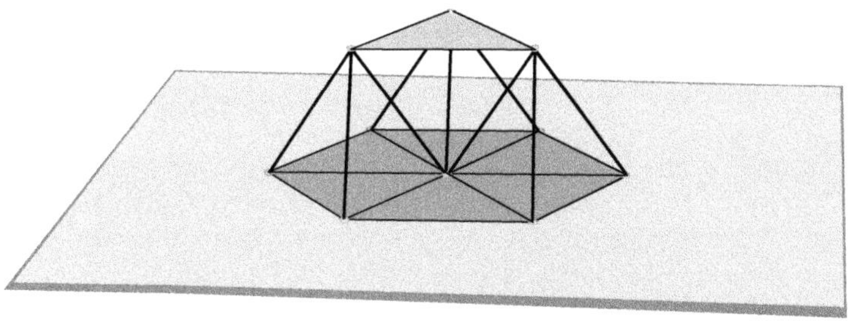

**FIGURE 8.5** Regular Tetrahedra Providing the Top of the Cupola

## An Example of a Task Only Existing in a Dynamic Geometry Environment

Students are given a plane and its equation in Cabri 3D (Figure 8.6). They can observe that, when moving a point defining the plane, this latter varies as well as its equation. A point *P* with displayed coordinates not belonging to the plane is given. Then the plane and all its points are hidden and students are asked to move point *P* in order to put it as close as possible to the hidden plane (Figure 8.7). Of course students must solve the task with the plane remaining hidden.

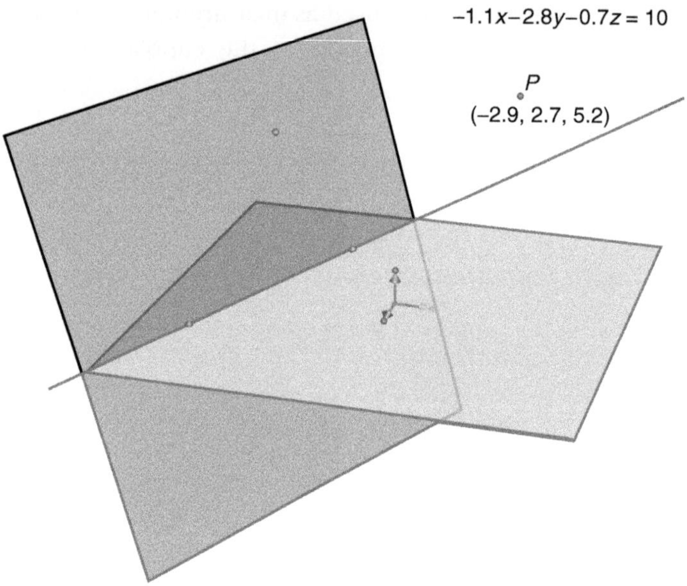


**FIGURE 8.6** Plane and its equation

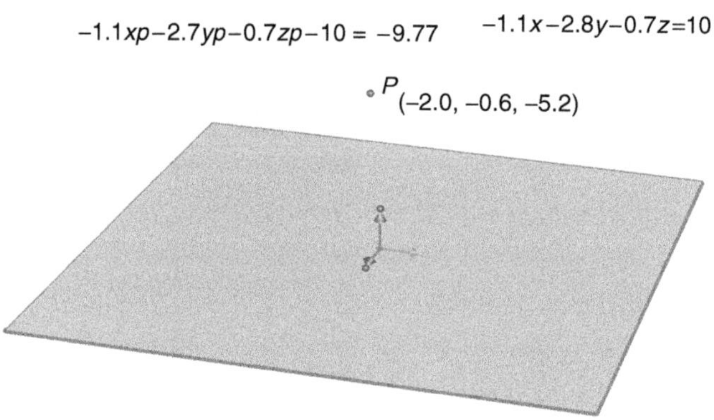


**FIGURE 8.7** The plane is hidden; point P is moved until the expression equals 0

Solving this task requires using the equation of the plane in absence of a spatial representation (Figure 8.7). Algebra is the only way to control the position of the point with respect to the plane. This is exactly why the task is interesting for understanding that the value of the equation instantiated with the value of the coordinates of P allows the students to know whether the point is on the plane, very close to the plane or not. Obviously this task can only be given in a dynamic environment in which geometric and algebraic representations are linked.

Such tasks given in dynamic geometry environments can contribute to learning for two reasons:

- the task itself and its demands
- the feedback given by the environment to the students' actions and in particular when dragging, the students may observe the effects of their construction in the environment, whether it is preserved by drag mode or not, whether it behaves as expected or not.

## Evolution of the Tasks of the Scenarios Over Time

Tasks of the first category were found only in the scenarios written by a teacher who was a novice at teaching but an expert in the use of technology: they were tasks in which Cabri was facilitating the collection of numerical data.

We also observed an evolution over time in the type of tasks proposed by experienced teachers. The first and second versions of their scenarios did not mainly comprise tasks that cannot be posed outside of the environment. It is easy to understand that the design of such tasks represents a conceptual break with the usual tasks performed in a paper and pencil environment. It is easier to discover new efficient strategies available in Cabri in an existing task than to invent new tasks. We could observe an evolution over time in the frequency of occurrence of the types of tasks. At the beginning, most were observation tasks for conjecturing, whereas more different tasks appeared in later versions.

From the very beginning of the writing of scenarios, the experienced teachers in both domains teaching and Cabri often used the observation tasks in which Cabri facilitated the production of conjectures. They expressed clearly that for them the role of the software is to save time, to avoid complex constructions requiring the use of properties that are exactly the properties to be discovered and to favour visualization. The declared intention of the teachers was to keep the demands of the task at a modest level. The role of Cabri was mainly to facilitate conjecturing and not to cause a problem, as in construction tasks, where the solving strategies have to be constructed with the Cabri tools.

At the beginning of the project, the experienced teachers also used construction tasks but less frequently. And finally, influenced by the researchers, they introduced a geometrical transformation as a black box task: students had to find out how the image of a point could be constructed from its pre-image.

## Multiple Dimensions Involved in the Design of Tasks

From this project, we could infer that the design of task resorting to dynamic geometry and aimed at fostering learning must be based on questions of a varied nature:

- epistemological questions: Is there a real problem to be solved? What kind of mathematical knowledge does require the solving of the task with technology? Is the most efficient solving strategy in the environment based on mathematical knowledge which is aimed by the task?
- cognitive questions: What kind of learning does promote the task? This analysis must be done by taking into account students former knowledge and conceptions.
- didactic questions: What are the means of action provided by the environment for solving the task? Are the values of the variables of the task chosen in order to promote the desired strategies? Is there feedback from the environment for invalidating wrong strategies? This is particularly important in dynamic geometry environments in which dragging offers feedback.
- instrumental questions: What do students know about how to use the environment to solve the task? Will their mathematical knowledge allow them to solve the task by using tools of the environment they are not familiar with or conversely can they build a new solving strategy capitalizing on their familiarity with the environment?

## Conclusion

These questions are presented separately for clarity reasons but they are clearly intertwined. Assude (2007) calls instrumental integration the way instrumental and mathematical dimensions are organized and related to each other by the teacher when giving tasks to students. From observations of teachers, she concludes to the high level of expertise required by coordinating both mathematical and instrumental dimensions. Such an expertise is far from being spontaneous and pre-service as well as in-service teacher education can certainly play a critical role in the development of such an expertise.

## Notes

[1] At the address sylvester.math.nthu.edu.tw/ d2/talk-atcm2006-unmotivated/
[2] Asian Technology Conference in Mathematics, http://landau.ma.polyu.edu.hk/atcm/

## References


Assude T. (2007). 'Teachers's practices and degree of ICT integration'. In Proceedings of CERME 5 (5th Conference of the European Society for Research in Mathematics Education) 1339–1348. http;//ermeweb.free.fr/Cerme5.pdf (accessed on 10 November 2010)

Brousseau, G. (1997). *Theory of Didactical Situations in Mathematics.* N. Balacheff, M. Cooper, R. Sutherland and V. Warfield (trans and eds). Dordrecht, The Netherlands: Kluwer Academic Publishers.

Laborde, C. (2001) 'Integration of technology in the design of geometry tasks with Cabri-geometry'. *International Journal of Computers for Mathematical Learning* 6, 283–317.

Chapter 9

# Learning from Acting on Objects

John Mason
*Open University and University of Oxford, UK*

## Introduction

This chapter brings a collection of psychological and pedagogical constructs to bear on the use of physical and virtual objects with which learners can interact in order to learn mathematics. Plato commended Egyptian use of simple apparatus for learning about number, and ever since the first HMI reports in the mid-nineteenth century, people have advocated use of pedagogic apparatus in secondary schools as well as in primary schools, but with little success. This is entirely consistent with a modern perspective of structural relations between actions on objects as being the stuff that mathematics formalizes and studies. Consequently electronic screens appear to provide an even richer array of virtual objects through which to display phenomena and to enable learners to act. Applying elaborations of Bruner's three modes of (re)presentation (*enactive, iconic,* and *symbolic*) seen as three worlds in which learners act, which currently provides a strong theoretical basis for structuring teaching, and applying his work with colleagues on Vygotsky's notion of *scaffolding*, augmented by the equally necessary component of *fading*, leads to some sharp questions about whether opportunities to manipulate virtual objects can ever be sufficient to ensure learning.

This chapter suggests that in order to learn from analysing phenomena and acting on objects it is necessary to do more than experience those phenomena and those actions. To learn *from* the experience it is necessary to withdraw from the action and to become aware of the effects of those actions as phenomena, including considering which actions are more or less effective in different situations, and of how structural relationships can be perceived as properties that can apply in other situations.

## Background

Information and Communications Technology (ICT) brings powerful tools into the hands of anyone who cares to access them. Drag & Drop, Fill-Right and Fill-Down, and simply Drag-to-Vary are powerful actions that can only be imagined in physical situations or when staring at a single diagram or collection of symbols on paper. ICT could therefore be expected to transform not only the teaching and learning of mathematics, but the very nature of mathematics as it is experienced in school and university. But does it? Will it? The track record so far is at best patchy, and research in several countries suggests that mathematics teachers are often among the least committed users of ICT in classrooms.

On the one hand, we have *Mathematica* and Wolfram's New Science (Wolfram, 2002) as but one representative of increasingly sophisticated mathematical software, and we have experimental mathematics as promoted by Jonathan Borwein (Bailey et al., 2007)) among others. On the other hand, we have the conservative forces of curriculum policy makers and politicians, together with the massive inertia of the institution of school mathematics. This inertia is due to the effort required in interacting with learners in new ways and on new topics that may be unfamiliar both to teachers and to learners.

Mathematics, especially in school and university, can be seen as ways of explaining and predicting phenomena through appreciating underlying structural relationships. These phenomena can be chosen to expose relationships that require both scoping and justification: Under what conditions might they hold, and why must they hold? Whereas phenomena in the material world such as movements of doors and diggers, uses of circles as manhole covers, sagging of overhead cables, forces on Frisbees, rolling of paper cups on floors, trajectories of fountains, scaling of maps and a wealth of other multiplicative relationships can be very complicated to disentangle, virtual phenomena can be constructed to focus attention on specific structural relationships that can be captured mathematically.

At its heart, mathematics concerns actions on objects, in a spiral of action becoming object to be acted upon. Thus numbers arise from the act of counting, arithmetic acts on numbers; arithmetic operations are special cases of functions which are actions on numbers that give rise to extended numbers such as rationals, reals, complexes and beyond; actions on functions leads to derivatives and integrals, transforms and functionals, and so on, in an apparently endless cycle. Phenomena with which the individual can interact can be explored, and relevant mathematical concepts can be experienced, through performing actions on objects. The objects may be material (using Pythagoras to ensure a right-angle, using string to locate altitudes of triangles) but are more likely to be conceptual leading to arithmetic and algebra. Electronic screens offer the prospect of an ever-extending rich array of domains in which mathematical actions can be performed and experienced.

## Theoretical Underpinnings

Jerome Bruner (1966) famously proposed three modes of (re)presentation[1] in human experience, which he called *enactive, iconic* and *symbolic.* Enactive (re)presentation meant for him the use of physical objects, but it can usefully be extended to refer to anything which is confidently manipulable by learners. An *iconic* (re)presentation is for him a diagram or image that 'looks like' what it (re)presents, and so is likely to be correctly interpretable in a relatively wide community. By contrast a *symbolic* (re)presentation is purely conventional and independent of the thing being (re) presented. In other words, to make sense of a symbolic (re)presentation you need to be told what it signifies.

Of course the same object can be enactive for some people, iconic for others, and symbolic for yet others, or perhaps can be treated in any of these three modes. For example, to someone competent with algebra, an $x$ suggests an as-yet-unknown and is confidently manipulable and hence enactive, whereas for most people it is an icon for 'that topic called algebra that I never understood', and entirely symbolic in its abstruseness (Mason 1980). One consequence of this extension of Bruner's distinctions is that the form of (re)presentation depends on the learner. Thus, a secondary student competent at basic arithmetic treats numerals as objects, whereas most pre-school children experience them as symbols. When symbols become confidently manipulable and meaningful, they lose their 'symbolic' qualities and become enactive elements for further manipulation.

Like Plato, Bruner recommended getting children working enactively with physical objects. He went further though and recommended getting them to use picture-diagrams to (re)present those actions while the objects themselves might remain in view but out of reach, although if significant difficulties were encountered the apparatus could be retrieved; and finally, getting children to work with symbols standing for those objects, with access to mental images of actions performed previously and to physical enactments. These transitions together constitute a form of scaffolding, which is gradually withdrawn in order to encourage and foster movement to the symbolic (Brown et al., 1989, Love and Mason 1992).

In trying to support mathematics teaching at primary and secondary level, the Centre for Mathematics Education at the Open University (1981, see also Mason and Johnston-Wilder 2004) augmented Bruner's ideas in the form of two further frameworks or labels for distinctions:

Manipulating – Getting-a-sense-of – Articulating

was intended to act as a reminder that the point of manipulating anything was not simply to get answers, but to develop over time a sense of underlying structural relationships which, when gradually brought to articulation, and refined could become confidently manipulable objects in their own right. This trio was seen as a form of helix so that when some difficulty arises it is possible to backtrack down the helix to something more confidence inspiring so as to then make sense of the situation. Once confidently manipulable objects are involved, learners can work their way back up the spiral at a pace and in a manner suitable to them. The helix was combined with

Do – Talk – Record

as a reminder not to push learners too quickly to written records before they have had a chance to do things (enactively) and to talk about what they were doing, not just as an informal chat but involving increasingly succinct and efficient use of technical terms. It was also pointed out that talking can inform doing as well as recording, and attempting to record can inform both doing and talking about what is being done. These were related to mathematical processes or powers that learners possess based on distinctions made by George Pólya (1962; see also Mason et al., 1982).

Finally, the notion of scaffolding and fading was independently reconstructed as

Directed – Prompted – Spontaneous

as a reminder of the importance of moving from giving learners specific instructions as to what to do, to increasingly indirect prompts so that learners begin to integrate the actions into their available functioning so that eventually they are able to act spontaneously *for themselves*. This instantiates Vygotsky's notion of the *Zone of Proximal Development* which is the collection of actions that learners can already carry out when cued, and are on the edge of being able to initiate and carry out for themselves (van de Veer and Valsiner 1991, Mason et al., 2007).

Bruner's trio provides the basic theoretical justification for many current pedagogical practices, particularly in primary school, while the additional frameworks, when internalized by teachers in any phase of education, provide reminders of choices of action and justifications for those actions when planning for or in the midst of a lesson.

## Examples

I shall consider two examples of mathematical phenomena to illustrate the issues arising from the use of ICT in classrooms. Neither are officially part of the school curriculum, yet the first is accessible to upper primary and secondary pupils as a phenomenon, and the second is accessible for exploring.

### Example one: Actions on a Line

I consider this to be one of the most amazing geometrical phenomena, certainly among those involving only straight lines.

Take any three points *A*, *B* and *C* on a line, (for the time being take *C* not to be the mid-point of *AB*). Now let *P* be *any* point in the plane not on the line *ABC*. Denote by *Pc* any point other than *P* or *C* on the line *PC*. Form the lines *A Pc* and *B Pc* and where they cross the lines *PB* and *PA* respectively, denote the points by *Pa* and *Pb*. Finally, let *D* be the point where the line *Pa Pb* meets the line through *ABC*. The amazing fact is that *D* is independent of the choice of *P* and the choice of *Pc*.

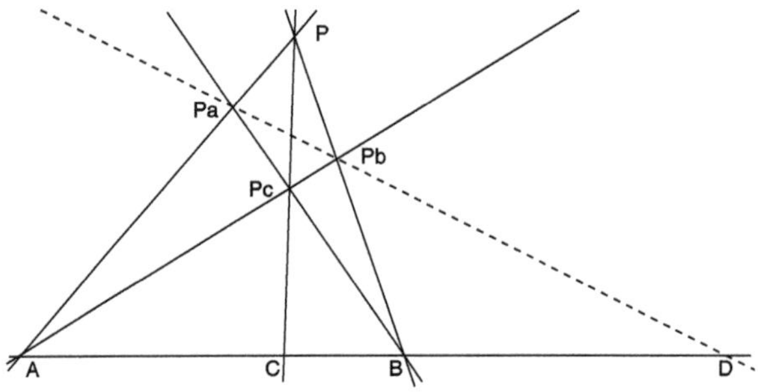


The principal phenomenon of interest is the wide range of freedom of the points *P* and *Pc* and yet the point *D* remains invariant. This can be experienced using dynamic geometry software in which *P* and *Pc* can be dragged about the plane while the point *D* fails to move, and that this holds for any relative positions of *A*, *B*, and *C* (when *C* is the mid-point of *AB*, the line *Pa*

*Pb* is parallel to the line *AB*). The enactive aspect is the dragging of points; the structural relationship is the invariance of the point *D*. It turns out that there is a special relationship concerning signed distances along the line:

$$\frac{\frac{AC}{CB}}{\frac{AD}{DB}} = -1$$

which is a ratio of ratios. It is intriguing that a relationship concerning four distances along a line should be manifested by intersections of lines in a higher dimension.

There are further relationships without even involving the point *D*. For example, considering ratios of lengths of segments, it turns out that

$$\frac{PP_a}{P_aA} + \frac{PP_b}{P_bB} = \frac{PP_c}{P_cC}$$

which provides a geometric method for adding ratios. Also

$$\frac{1}{CP_c} = \frac{1}{AP_a}\frac{\sin(AP_cC)}{\sin(AP_aB)} + \frac{1}{BP_b}\frac{\sin(BP_cC)}{\sin(BP_bA)}$$

which, when *P* goes to infinity in any direction, yields the harmonic sum

$$\frac{1}{CP_c} = \frac{1}{AP_a} + \frac{1}{BP_b}$$

associated with the crossed ladders and two-courier problems of medieval arithmetic (Dagomari 1339, see Smith 1908). For example,

> From Noemberg to Rome are 140 miles: A Traveller sets out at the same Time from each of the two Cities, one goes 8 Miles a Day, the other 6: In how many Days from their first setting out will they meet one another, and how many Miles did each of them go? (Hill 1745, p. 365)

Graphing the two journeys, and then rotating the graph through 90° produces a 'crossed-ladders' diagram which has the property that the sum of the reciprocals of the heights of the two ladders is the reciprocal of the height where they cross.

Here the enactive elements are much harder to 'see' and manipulate, and very hard to experience bodily because they involve multiplicative relationships. The ratio of two ratios is hard to experience enactively, until the diagram becomes a confidently manipulable entity for expressing the ratio of ratios! Use of measurements to indicate invariant ratios is much less convincing than physical embodiment, based as they are on approximate measurements by computer. The third relationship admittedly uses trigonometry, but the connection to classic word-problems is remarkable.

### Example 2: Polydials

Polydials is software still under development, but it illustrates some of the potential as well as some of the obstacles of pedagogic software. Available to the user are dials with various alterable whole numbers of 'hours' marked on them. Also available are buttons used to advance the hands by a single 'hour' of one or more clocks simultaneously. Pictured below is a pair of clocks, one

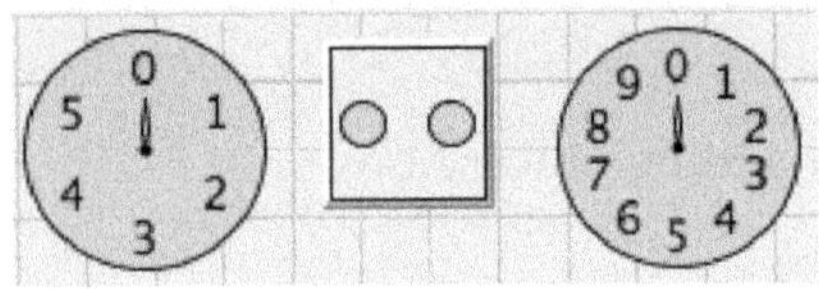


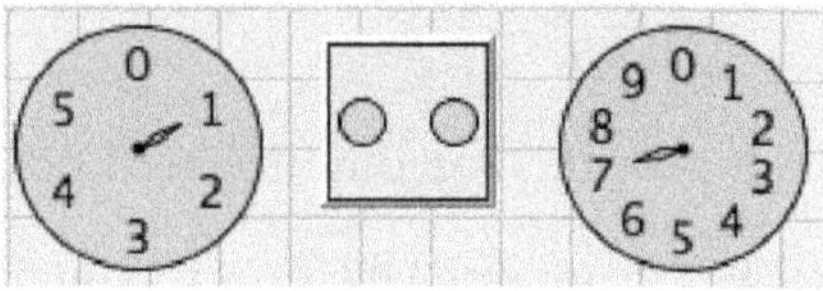


with 6 'hours' and one with 10, together with a button that advances both simultaneously. The second figure shows the result of 7 clicks
An immediate question is how many clicks of the button will be required to get both the dials back to the 0 position again, and what relative positions are possible for the two dials. This question in some sense abstracts numerous situations in which two things are happening at different rates, such as two wheels of different diameters on a vehicle, or two people walking at different rates or using different pace-lengths and so on.

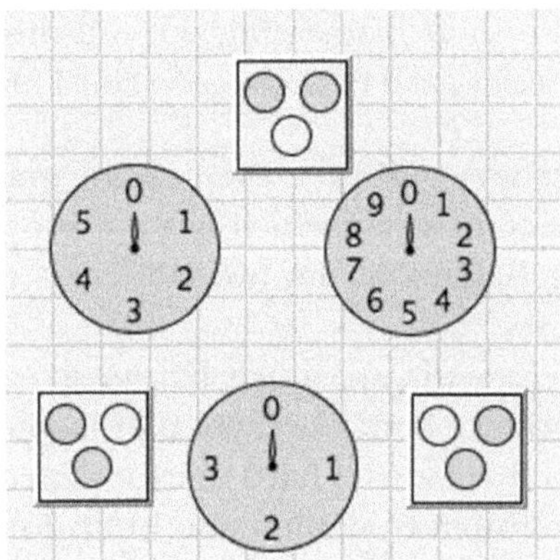

A more challenging situation arises when there are more dials and more buttons.

One task might be to try to get all the dials reading the same specific 'hour' such as 1. An alternative version is to start with the dials in some non-zero positions and then try to get them all back to 0 again simultaneously. Each button advances only the shaded dials in its icon.

It is worth noting that the button 'labels' are icons chosen to indicate 'what they do'. The question is then how many pushes of each will achieve some stated goal. Now there are two possibilities: empirical exploration (by pushing buttons, at first almost at random, then increasingly systematically or with structural intent) and theoretical analysis (by expressing underlying structural relationships as properties).

Let the button icons also be used to stand for the number of pushes required to reach a particular goal, say all dials pointing to 1. Then because the order in which you push buttons to advance the hands does not matter, the following 'equations' express the desired relationships.

 +  $= 1 + 6s$ where $s$ is some integer,

since these are the two buttons that effect the 6-dial, and any extra multiple of 6 pushes leaves the dial in the same place.

 +  $= 1 + 4t$ where $t$ is some integer

 +  $= 1 + 10u$ where $u$ is some integer

The icons can themselves stand for the number of times they have been pushed, thought of as a variable or as an as-yet-unknown. The equations can readily be seen to be unsolvable, because

 –  $= 6s - 4t$ which is even, whereas  +  $= 1 + 10u$ which is odd.

So that task is impossible. If instead of 1, the target was to get the dials pointing to 1, 2, and 5 respectively, then the equations are

 +  $= 1 + 6s$ where $s$ is some integer,

 +  $= 5 + 10u$ where $u$ is some integer

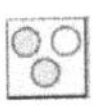 +  $= 2 + 4t$ where $t$ is some integer

and these can be solved, to give

2  + 2  + 2  $= 8 + 6s + 4t + 10u$ (adding all the equations).

Dividing by two and subtracting the equations in turn gives

 $= 2 + 3s - 2t + 5u$;  $= 3 - 3s + 2t + 5u$;  $= -1 + 3s + 2t - 5u$;

To get a non-negative number of pushes for each button, take $s = 1$, $t = 0$, $u = 0$, yielding 5, 0 and 2 pushes respectively.

Here the icon for the button is used 'symbolically' to express desired relationships. If users have done a good deal of button pushing and resetting, they may be glad of a structural rather than empirical approach to deciding whether some goal is in fact possible at all.

The software offers extra 'support' for checking relationships, by treating each button as an action, and multiplying it by the number of times you want it to be pressed.

The mixture of icons and coefficients  becomes an expression for the combined action, which can then itself become a button. In this way learners are 'actually' operating in a module over the integers, but their experience is of combining numbers of button pushes.

There are two pedagogic difficulties emerging from Polydials. One is the mathematically powerful switch of meaning of an icon (or a symbol) from an action to counting the number of times the action is performed. This difficulty is highlighted in ICT because of the difference between using a name as an object to manipulate, and using a name to refer to something else like a value. The other difficulty is endemic to software and tool development generally. At every stage of sophistication, it quickly becomes evident that 'it would be nice to be able also to . . . '. In the case of geometry, such software would be able to suggest extra constructions, or the user could select some elements of a diagram and treat them as a relationship to be combined with other geometric relationships; in the case of Polydials it is the desire to have access to some sort of 'icon processor' that, given the icon-equations, will solve them in some perspicuous manner. Of course symbol processing software such as *mathematica* and *maple* do this, but they require significant algebraic competence on the part of the user. The point is that it is highly likely that there will always be 'another layer' of generality of action that 'it would be nice' if software could perform perspicuously.

## Theoretical Difficulties and Practical Obstacles

There are differences between experiencing phenomena, being moved to explain or justify them, and using tools to suggest, if not provide, that explanation or justification. For example, dynamic geometry software can convince you that some property always holds without any hint of why it must be the case. In the first example, it is not at all clear from dragging why it is that point *D* necessarily remains invariant. Indeed it requires either trigonometry or linear algebra to be fully convinced of the necessity. Very often in geometry, what is required is the addition of one or more extra 'construction' lines. In example 1 a perpendicular from *P* to the line *AB* opens up a number of right-angled triangles and hence trigonometric relationships that can be expressed and manipulated.

Once you have a proof, it is often possible to build an animation which shows the steps in the reasoning by directing attention to appropriate elements through the use of colour, thickness and flashing. Most of the many different proofs of Pythagoras' theorem can be displayed in this way as phenomena to be made sense of, as can many other theorems. The learner is invited to describe, to bring to articulation the 'animation sequence', and in the process, to re-construct the proof. However, it will only be a proof for them if they are simultaneously aware of what in the particular diagram can be changed and what must remain invariant, and how these invariances are called upon in the reasoning.

Building up a proof through combining steps in reasoning suggested by an animation or sequence of frames may support learners in experiencing complex reasoning; getting them to be able to do the same for themselves requires multiple shifts of attention as they discern pertinent elements, recognize structural relationships between those elements, introduce new elements so as to be able to express and connect relationships, and perceive those relationships as properties applying to a whole class of figures or objects and not simply the ones in a diagram or in a particular case. Only then does it make sense to isolate particular properties and reason solely on the basis of those alone.

Even though facts, including statements of theorems and justifications for those theorems, are readily available on the web, to use them as tools in your own reasoning requires more than access to the internet. It requires sophisticated mathematical thinking, including awareness that there could be a relevant theorem, making connections through recognizing chains of relationships, and competence, disposition and perseverance to pursue and check the reasoning needed for justifying any new or newly encountered phenomenon.

## Learning from the Past

As mentioned in the Introduction, engaging learners by getting them to interact with objects has an ancient pedigree: Plato praised Egyptian schools for their use around 400BCE (Hamilton and Cairns 1961 p. 1388–9). But while the use of apparatus for teaching primary mathematics is well developed, it is not at all clear that it is always effective. Experience alone is not enough to guarantee learning from that experience. Use of pedagogic apparatus does not have a good track record at secondary. The HMI reports in the UK have frequently noted the absence of apparatus, and attempts to promote the use of apparatus in secondary have been at best marginally successful. For example, Cuisenaire rods became very popular for a time while there was an extensive infrastructure (Goutard 1974) but together with Dienes Blocks or Multi-Base Blocks they can be found at the back of many school cupboards.

The problem is that despite the rhetoric, apparatus does not embody mathematics. Rather, mathematics can be *seen as* embodied in apparatus. Sometimes it is not clear whether you have already to understand something in order to perceive the embodiment! The subtle shift is at the heart of the difficulties: it is not the apparatus that teaches, but rather the way of working with the apparatus.

Following Bruner's elaboration of his three modes of (re)presentation, Kath Hart (1993) mounted a campaign for a time trying to find evidence for the effect of the use of enactive apparatus in secondary classrooms, and came to the conclusion that the most common practice was to get learners using apparatus, and then to switch to symbolic expression, without much attention to the way in which the symbols express the actions performed with the apparatus. Whereas Dienes (1960) and others had thought in terms of mathematics being embodied in physical objects, a chorus of counter-claims grew, pointing out that the mathematics is in the eye of the beholder, not in the apparatus, nor even in the phenomenon.

Learners almost always need to have their attention directed to the mathematical aspects of the actions performed on the objects. Indeed, mathematical concepts emerge as the expression of the structure of those actions, perhaps as delineated by Bruner (1966) and perhaps in other ways as well.

The following quotations trace some of the controversy over the pedagogic use of apparatus.

> One of the problems about using symbolism is how to find the best time for introducing it. If this is done too early, it tends to be an empty shell. Classroom work in mathematics can so easily degenerate into learning certain rules by which the signs can be manipulated, and studying situations in which they are applicable, each application being separately learned. This of course is necessary if the signs do not symbolize anything. On the other hand, it is possible to wait too long before introducing symbolism. When a child has become familiar with a mathematical structure he needs a language in which to talk about it, think about it, and eventually transform it. New constructions need new names, their properties must be described by new symbols if more of the detail of the structure is to be grasped at one time, and so reflected upon more effectively. (Dienes 1960 p. 160–61)

Links to the Open University frameworks are perfectly evident in Dienes' analysis.

> Before we recommend to teachers that they use manipulatives we should advise them to view the appropriateness and limitations of the materials for the purpose of leading to and authenticating a part of formal mathematics. [ . . . ] We need to research when manipulatives are appropriate as well as the balance of time given to different activities within the same scheme. (Hart, 1993, p. 27–28)

If learners' sense of a concept is limited to particular contexts in which they manipulate some physical objects, or if they associate a concept with particular objects, then the experience is as likely to be an obstacle as a help. The same applies to metaphors such as 'an equation is a balance' which runs into difficulties if it is maintained when you want to subtract things from both sides (some people use 'helium balloons'). As Dienes indicates, it is as important to plan when to withdraw from reliance on the actions (and the metaphors). Similar difficulties can arise when expression such as 'larger from smaller you can't do' because some children refuse to alter an established practice based on such a falsehood.

The whole point of performing actions is to pay attention to the effects, to become aware of underlying structural relationships:

> In mathematics learning, the intention to make sense is essential (Erlwanger, 1973).

Educators have railed against the misuse of apparatus. For example:

> Often when manipulatives are used in teaching mathematics, the teacher demonstrates *the* way they are to be used and students are left little freedom to give meaning to the experience in ways that make sense to them; the way the materials are to be used is prescribed. There is the mistaken belief on the part of the teacher that the mathematics is apparent in the materials, for example, 'base 10' blocks (Cobb, Yackel and Wood, 1992). This is based on the belief that mathematics is 'out there' and that models 'show' the concepts. The demonstration with concrete materials is quite appealing because the concepts are so vivid for those who have *already* made the construction. Thus there is the mistaken belief that since we, as adults, can see the mathematics in the blocks, the students will too. But the 'seeing ' requires the very construction the activity is intended to teach. (Wheatley 1992, p. 534)

Piaget emphasized the necessity of learners constructing meaning for themselves (Piaget 1980, p. 90–91), and Vygotsky (1978) emphasized the importance of the social practices within which the instruction is embedded as a vital component of sense making.

> Manipulatives are not, of themselves, carriers of meaning for insight. 'Although kinaesthetic experience can enhance perception and thinking, understanding does not travel through the finger tips and up the arm' (Ball, 1992, p. 47). It is through their use as tools that students have the opportunity to gain insight into their experience with them. Research has shown that for children to use concrete representation effectively without increased demands on their processing capacity, they must know the materials well enough to use them automatically (Boulton-Lewis, 1998). If the user is constantly aware of the artifact then it is not a tool, for it is not serving the purpose of enabling some desired activity which moves one toward a desired goal state (Winograd and Flores, 1986). . . . Students sometimes learn to use manipulatives in a rote manner, with little or no learning of the mathematical concepts behind the procedures (Hiebert and Wearne, 1992) and the inability to link their actions with manipulatives to abstract symbols (Thompson and Thompson, 1990). This is because the manipulative is simply the manufacturer's representation of a mathematical concept that may be used for different purposes in various contexts with varying degrees of 'transparency'. (Moyer, 2001, p. 176–77).

Algebraic symbols do not speak for themselves. What one actually sees in them depends on the requirements of a specific problem to which they are applied. Not less important, it depends on what one is *prepared* to notice and *able* to perceive. (Sfard, 1994, p. 192)

In order to promote effective learning from experience, Simon et al. (2004) emphasize the importance of reflections on the effect of actions. It is not simply a latter of getting learners to act on objects, but to get them to become aware of the effects of those actions. Furthermore, people develop preferences for particular ways of acting which may not be either efficient or readily abstracted. Anne Watson refers to super-methods as those which apply to a wide variety of situations rather than being limited to a few special cases. For example, to find the area of a rectangle there are many different ways; which of these extend to other rectilinear shapes or to other quadrilaterals? Counting squares is much less efficient than having a formula based on the lengths of the edges and the measure of sufficiently many angles.

## From the Physical to the Virtual: But How Do You Get Back?

Electronic screens seem to provide a domain with enormous potential for replaying all the confusions and misapprehensions about the role of physical apparatus in teaching and learning mathematics. They also present some additional obstacles, as well as providing new affordances.

Merely manipulating objects on a screen is likely to be analogous to manipulating physical objects: it is tempting for teachers to instruct learners in what to do, to get them to rehearse those actions, and then to expect learning to come about. Even where the instruction includes careful introduction of technical language for describing and guiding those actions, learners are just as likely to absorb the practices without being able to generate those actions for themselves from understanding as in any other form of instruction. The *didactic tension*, a consequence of the *didactic contract* of Brousseau (1997) is ever-present:

> The more clearly the teacher indicates the behaviour being sought, the easier it is for learners to display that behaviour without generating it from themselves.

Because of the *didactic contract*, learners exert pressure on teachers to 'tell them what to do', and sometimes, due to the apparent shortage of time, it even appears to be efficient. But the whole point of manipulating objects is to become aware of, to get-a-sense-of, and then to try to bring to articulation, relations between objects, and relations between actions. The extra difficulty presented by objects of screens is that there is a curious phenomenon associated with closing down the software or turning off the machine: the images, the bodily awarenesses evaporate so suddenly that there is often a minute gap when nothing at all is being attended to (Mason 1985). Thus learning from virtual objects depends on an ongoing process of sense-making, of trying to articulate that sense, and of creating labels for actions that prove to be effective. The speed of computers makes empirical guess-and-test more viable and more attractive than ever before, yet it is precisely the awareness that structural relationships can be used to generate efficient techniques for resolving problems and for explaining or predicting phenomena that marks significant learning.

The constructs of *situated cognition* and *situated abstraction* have been used to act as reminders that people can readily learn to carry out specific actions in specific situations, but recognizing something about a new situation that brings to mind actions carried out in the past, and reconstructing those actions from understanding of the structural relationships that makes them effective requires much more. When you return to some software, like riding a bicycle after a period away, a great deal of locally achieved competence can come flooding back. But the issue is whether what is learned can be used elsewhere in the future. How do you take something useful from a virtual world and use it in the world of mental imagination, or in the world of material objects? What is it exactly that is useful? All this requires ongoing reflection and re-construction.

## Conclusion

It is tempting to conjecture that as virtual objects become more sophisticated, and as the actions that can be performed on and with them become more complex, the necessity for disciplined ways of working involving withdrawing from action and bringing to articulation what the actions are, what their effects are, and in what circumstances might those actions be useful in the future can only increase markedly. To emerge from the virtual with actions and awareness of relationships that can be used effectively and efficiently in the material and mental worlds requires disciplined practices. These are what really need to be taught.

## Note

[1] I choose to put brackets around the prefix because in most cases what is being offered is a presentation rather than a re-presentation, and I find it very helpful to be re-minded of this when thinking about actions on objects.

## References

Bailey, D., Borwein, J., Calkin, N., Girgenshohn, R., Luke, D. and Moll, V. (2007). *Experimental Mathematics in Action.* Wellesley: A. K. Peters.

Ball, D. (1992). 'Magical Hopes: Manipulatives and the Reform of Math Education'. *American Educator,* 16 (2), 14–18, 46–7.

Boulton-Lewis, G. (1998). 'Children's strategy use and interpretations of mathematical representations'. *Journal of Mathematical Behavior,* 17 (2), 219–37.

Brousseau, G. (1997). *Theory of Didactical Situations in Mathematics: Didactiques des Mathématiques, 1970–1990.* N. Balacheff, M. Cooper, R. Sutherland and V. Warfield, (trans), Dordrecht, The Netherlands: Kluwer Academic Publishers.

Brown S., Collins A. and Duguid P. (1989). 'Situated cognition and the culture of learning'. *Educational Researcher,* 18 (1), 32–41.

Bruner, J. (1966). *Towards a Theory of Instruction.* Cambridge, MA: Harvard University Press.

Cobb, P., Yackel, E. and Wood, T. (1992). 'Interaction and learning in mathematics situations'. *Educational Studies in Mathematics,* 23, 99–122.

Dagomari, P. (1339). *Trattato d'Abbaco. D'Astronomia, e di Segreti Naturali e medicinali.* ms.

Dienes, Z. (1960). *Building up Mathematics.* London: Hutchinson Educational.

Erlwanger, S. (1973). 'Benny's conception of rules and answers in IPI Mathematics'. *Journal of Mathematical Behaviour* 1 (2), 7–26.

Hamilton, E. and Cairns, H. (eds) (1961). (W. Guthrie, Trans.), *Plato: The Collected Dialogues Including the Letters,* Bollingen Series LXXI. Princeton, NJ: Princeton University Press.

Hart, K. (1993). 'Confidence in success'. In I. Hirabayashi, N. Nohda, K. Shigematsu and F-L. Lin (eds) Proceedings of Psychology of Mathematics Education, PMEX-VII, Vol.1, p. 17–31, University of Tsukuba, Tsukuba,.

Hiebert, J. and Wearne, D. (1992). 'Links between teaching and learning place value with understanding in first grade'. *Journal for Research in Mathematics Education,* 23, 98–122.

Hill, J. (1745). *Arithmetick, Both in the Theory and Practice, Made Plain and Easy in All the Common and Useful Rules, Both in Numbers and Fractions, Vulgar and Decimal . . . the Like Not Extant.* Printed for R. Ware, J. and P. Knapton, T. Longman, C. Hitch and L. Hawes, J. Hodges, S. Austen, J. and J. Rivington, J. Ward, and M. Cooper, London.

Love, E. and Mason, J. (1992). *Teaching Mathematics: Action and Awareness.* Milton Keynes: Open University.

Mason, J. (1980). 'When is a Symbol Symbolic?' *For the Learning of Mathematics* 1 (2), 8–12.

Mason, J. Burton, L. and Stacey, K. (1982). *Thinking Mathematically.* London: Addison Wesley.

—(1985). 'What do you do when you turn off the machine?' Preparatory paper for ICMI conference: *The Influence of Computers and Informatics on Mathematics and its Teaching,* pp. 251–56. Strasbourg: Inst. de Recherche Sur L'Enseignement des Mathematiques.

Mason, J., Drury, H. and Bills, E. (2007). 'Explorations in the zone of proximal awareness'. In J. Watson and K. Beswick (eds) *Mathematics: Essential Research, Essential Practice: Proceedings of the 30th annual conference of the Mathematics Education Research Group of Australasia,* vol. 1, pp. 42–58. Adelaide: MERGA.

Mason, J. and Johnston-Wilder, S. (2004). *Designing and Using Mathematical Tasks.* Milton Keynes: Open University. Re-published (2006). St Albans: Tarquin.

Moyer, P. (2001). 'Are we having fun yet?: How teachers use manipulatives to teach mathematics'. *Educational Studies in Mathematics,* 47, 175–97.

Piaget, J. (1980). *Adaptation and Intelligence: Organic Selections and Phenocopy.* Chicago: University of Chicago Press.

Polya, G. (1962). *Mathematical Discovery: On Understanding, Learning, and Teaching Problem Solving.* Combined edition. New York: Wiley.

Sfard, A. (1994). 'Reification as a birth of a metaphor'. *For the Learning of Mathematics,* 14 (1), 44–55.

Simon, M. A., Tzur, R., Heinz, K. and Kinzel, M. (2004). 'Explicating a mechanism for conceptual learning: Elaborating the construct of reflective abstraction'. *Journal for Research in Mathematics Education.* 35 (5), 305–29.

Smith, D. (1908). *Rara Arithmetica: A Catalogue of the Arithmetics Written before the Year MDCI with a Description of Those in the Library of George Arthur Plimpton of New York by David Eugene Smith of Teachers College Columbia University.* New York: Athanaeum Press.

Thompson, P., and Thompson, A. (1990). 'Salient aspects of experience with concrete manipulatives'. In Proceedings of the 14th Annual Meeting of the International Group for the Psychology of Mathematics, pp. 46–52. Mexico City

van der Veer, R. and Valsiner, J. (1991). *Understanding Vygotsky.* London: Blackwell.

Vygotsky L. (1978). *Mind in Society: The Development of the Higher Psychological Processes.* London: Harvard University Press.

Wheatley, G. (1992). 'The Role of Reflection in Mathematics Learning'. *Educational Studies in Mathematics,* 23, 529–41.

Winograd, T. and Flores, F. (1986). *Understanding Computers and Cognition: A New Foundation for Design.* Norwood: Ablex,

Wolfram, S. (2002). *A New Kind of Science.* www.wolframscience.com (accessed on September 2009).

Chapter 10

# A Case Study of Using Multiple Resources to Teach Straight Line Graphs

Rosemary Deaney and Sara Hennessy
*Faculty of Education, University of Cambridge, UK*

## Introduction

We present the findings of a video-based case study of teaching and learning about straight line graphs. The research was part of the wider T-MEDIA project[1] (Hennessy and Deaney, 2007) which investigated how teachers promote student learning in secondary school subject lessons incorporating the use of Information and Communications Technology (ICT), particularly digital data projection and interactive whiteboard systems. This account focuses on how an experienced UK mathematics teacher, Sarah, exploited a variety of digital tools over a six-lesson series to develop her 12- to 13-year-old students' understanding of the concepts of intercept and gradient of straight line graphs. Through collaborative review of lesson videos with Sarah and her departmental colleague, we identified and documented key pedagogical aspects of her approach (see Hennessy and Deaney, 2009 for a detailed account of this process).

Sarah was head of her mathematics department and committed to working with mixed groups of students across the attainment range. She had participated in previous research with us and we knew that technology use was embedded within her practice. She routinely used a data projector and half class set of laptops in her lessons, with a variety of software; in the observed lessons this included dynamic graphing (Autograph) and spreadsheet packages, and an online Mathematics site (MyMaths).

Our analysis highlighted Sarah's pedagogical strategies for mediating learning with technology within four broad areas[2]:

1. Using multiple tools and resources
2. Fostering a supportive and collaborative learning environment

3. Developing concepts and responding to learning needs
4. Capitalizing on unexpected outcomes and errors

## Using Multiple Tools and Resources

Sarah believed that using technology made it easier to vary activities and segment work in ways that were motivating to pupils. She favoured 'mixing and matching' materials from different sources, rather than adopting particular programmes. The department had no interactive whiteboards but Sarah used a tablet computer linked with the data projector which also acted as a mobile input device for pupils during whole class teaching. Throughout the series, Sarah brought into play a variety of software tools selected for different purposes, including: online games to practise coordinates; spreadsheets to generate a formula to calculate the incremental increase of a given variable and produce a corresponding line graph and Autograph to generate lines from given equations. There were also more open-ended explorations of linear equations using an interactive tool in MyMaths, and use of a class response system to assess learning via teacher-generated multiple choice items. Non-ICT resources included matched cards for graph recognition and off-the-shelf and custom worksheets used in conjunction with tasks on the computer.

## Fostering a Supportive and Collaborative Learning Environment

One of the most notable aspects of Sarah's practice was the highly supportive and co-operative learning environment through which she *encouraged pupil participation.* Activities and interventions were carefully structured to offer appropriate challenge or support for individual learners, developing *confidence* to participate.

Sarah consistently *gave status* to, and *valued* pupil contributions, drawing on these as a resource. There was a deliberate strategy of praise and encouragement, especially through focusing on the correct or relevant part of a pupil's response.

*Collaboration* was fostered in both whole class and group-work situations. Sarah built on pupils' answers by further questioning or sensitively drawing in other pupils to challenge, corroborate or offer assistance. If a pupil became 'stuck' for an answer, they were invited to call on a peer. Where solutions were gained collaboratively, Sarah's feedback was inclusive, *crediting peer aid.*

Sarah often called pupils up to the whiteboard to explain some aspect, seeing this as beneficial to the learning of all concerned. Strategic questioning as she circulated prompted pupils to formulate and articulate *(rehearse)* their thinking, making it easier to speak out in whole class discussion when invited later on.

Watching and monitoring peer input at the board was a powerful mechanism for engaging pupils through *vicarious involvement.*

> If they make the wrong move, or they're about to do something, other pupils will actually join in and tell them what to do . . . which they wouldn't do if it was me.

This was extended through pupil use of the tablet; for example, after plotting a line parallel to $y = 2x + 3$, Sarah challenged one pair to enter a new parallel line using the tablet, thereby demonstrating their knowledge to the class.

Weaker pupils were often paired with more able ones, but required to fill in worksheets individually. *Peer tutoring and direction* were encouraged and grouped arrangement of tables facilitated this style of working.

### Motivating Pupils through Making Activities Accessible

Sarah offered accessible, *focused* learning activities and *reduced writing requirements*; for example, providing a spreadsheet template helped pupils to focus on 'seeing a relationship' rather than on the routine task of generating a graph.

Whole class games were used to introduce or revise key terms (intercept, origin, gradient) and (MyMaths) online games were a 'fun way to practise co-ordinates' in pairs. Sarah held that students '*learn by doing*, not just by watching' but her interventions were influential in 'mathematizing' situations in relation to lesson aims.

## Developing Concepts and Responding to Learning Needs

Tasks were designed to mobilize or scaffold *discovery* of the relationship between the equation and its graph:

> To be able to use Autograph and just plot $y = x + 2$, $x + 3$, $x + 4$, $x + 50$, $x - 2$, they are going to be able to see straight away that whatever num-

> ber you add or subtract at the end of the equation is the intercept on the y-axis . . .

The technology enabled her to *focus* on particular aspects such as the gradient and the intercept and then to discuss how they 'fitted in'.

Sarah provided highly responsive assistance. Sensitivity to group and individual needs led her to change course flexibly where needed, directing pupils towards targeted materials. Sarah was also mindful of the *constraints* as well as the benefits of available resources. For example, guarding against pupils 'just zooming through' online screens without engaging sufficiently with content, she circulated and monitored activity, questioning to check understanding.

Sarah's interactions were carefully phrased to elicit knowledge, make reasoning explicit – subtly *'reshaping' thinking* or guiding in a particular direction through *'funnelling'* prompts. As well as scaffolding activity, Sarah also strategically withdrew support (*'fading'*), stimulating pupils to take the next steps themselves.

### 'Intertwining' resources and planning activities to scaffold learning

Teaching offered staged introduction to new concepts through referencing everyday objects (e.g., discussion of 'slope'). It was interspersed with *exploration, discovery* and practice (mainly using Autograph, guided with worksheets). Later on, the MyMaths interactive tutorial was used to consolidate knowledge – which was then applied in a final activity requiring generation of parallel lines.

Sarah drew on an array of digital and non-digital resources to hand in the classroom to illustrate her explanations, for example representing positive and negative numbers with coloured plastic cubes.

She also built in tasks to *prime* for forthcoming activities and *pre-empt conceptual difficulty*. For example a starter activity relating fractions with division preceded calculating gradient:

> I wanted them to see . . . the distinction between 4 over 2 and 2 over 4 because I thought that would be a point where they might get a little bit muddled – and sure enough, later on in the lesson, it was!

There was a *balance of 'independent' enquiry with collective knowledge building*, as Sarah switched between whole class and small group modes: for example generating a series of graphs led pupils to notice that 'the number before x was always the same as the gradient'; using worksheets to guide their investigation, pupils then plotted graphs manually from given equations and checked their work using Autograph; the final plenary explored the visual effects of altering the gradient. This guided approach helped to guard against the danger of pupils acquiring idiosyncratic knowledge (Godwin and Sutherland, 2004), becoming 'stuck' or changing too many parameters at once and so failing to see the effect of one (Goldenberg, 1995).

Activities were interspersed throughout lessons with *mini-plenaries*, where Sarah explained and discussed emerging issues or introduced, demonstrated or modelled the next step. Projection technology assisted here, especially for stepwise knowledge building, consolidation and assessment of learning.

Sarah felt that one of the dangers of using ICT was that 'pupils don't relate that to how they would do things on paper' so intertwining modes of working was designed to ensure they had 'both sets of skills and are able to see the links between them.' For this reason the second lesson comprised paper-based consolidation activities for topics already given an ICT-based treatment. As well as recapping previous knowledge, these activities helped to familiarize pupils with the relationship between an equation and its graph and to prime for investigative work using Autograph in subsequent lessons.

Sarah *blended use of ICT and non-ICT tools*. Visual display, both at the front of the class and on laptop screens, formed an important focus for dialogue and joint reference at whole class and small group levels. Projecting materials onto a rollerboard enabled them to be annotated with a marker pen. A separate, non-digital whiteboard was also used to display key reference material during the lesson, for example, aims, key words and software access instructions. It also provided an area where pupils could record or work through ideas during whole class interactions, as an adjunct to the main projected display.

## Capitalizing on Unexpected Outcomes and Errors

The school's ethos emphasized the value of learning from mistakes and Sarah's approach encouraged pupils to develop this perspective:

> Some of them have still got this idea that things are 'wrong' and 'right' in Maths and it's trying to get them away from that . . . and it doesn't matter if you're thinking along a different line – that could be useful as well.

The *instant feedback* afforded by technology was seen as helpful in recognizing errors and Sarah made space for pupils to *spot mistakes* and anomalies. For example, pupils were asked to suggest points on the line $y = 2x + 1$ beyond the segment visible on screen. One pupil's suggestion of (111, 0) was accepted provisionally, and written on the board. As the discussion progressed, he recognized his mistake and asked to correct it; Sarah's prompting and peer aid enabled him to reach the answer of (111, 55), a solution verified by show of hands from the group.

Technology was deliberately harnessed to facilitate *self-correction*, for example auto-plotting on the computer served to corroborate manual graph production and highlight plotting errors. Interestingly, use of technology itself generated some unexpected results (Hoyles and Noss, 2003). Sarah's mediation was pivotal in achieving learning gains when these situations arose. For example, during use of Autograph, a pupil was surprised and pleased to obtain two parallel lines on screen after having entered $y = -2x +$ and $-2$. Sarah explained the meaning of the $\pm$ symbol that had been generated by Autograph from his input, then prompted him to identify intercepts on the y-axis and to observe that the gradients of the two lines were the same. This example was particularly fortuitous as it meshed with the aims of the lesson.

## Conclusion: An Adaptive Approach to Harnessing Technology

These various facets of Sarah's approach collectively illustrate an *adaptive style of teaching* that embraces the wide diversity of individual differences encountered in a very mixed class of students. Rather than seeing these differences as obstacles, Sarah exploited them through designing activities and student pairings where learners with different profiles or styles could work together productively. Her adaptive teaching involved continuously offering support and challenge in direct response to learning needs and motivation – in order to create a learning arena in which everyone could participate with confidence.

## A Multimedia Resource for Collaborative Teacher Development

The four themes outlined above are described and illustrated in a multimedia resource produced by the research team (Figure 10.1).[3]

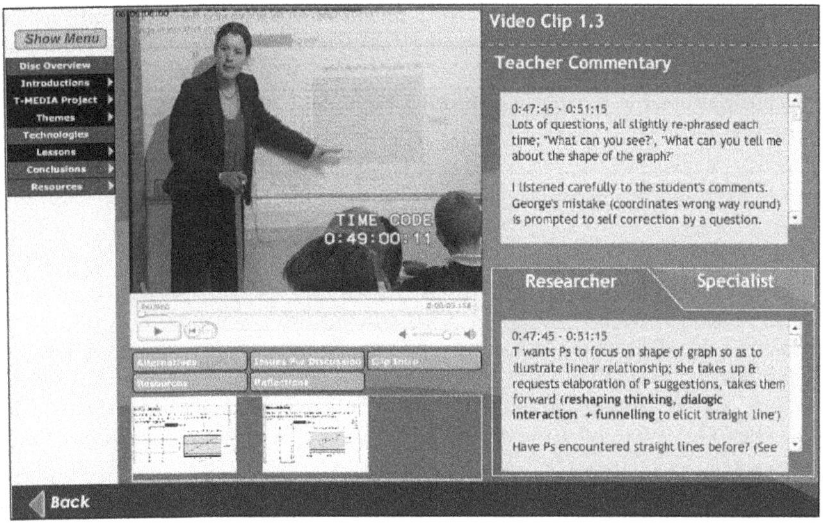


**Figure 10.1** A Video Clip Screen from the T-MEDIA Resource

It contains a set of hyperlinked video clips and analytic commentary, with built-in points for reflection and discussion. Designed as a tool for continuing professional development (CPD), the resource aims to stimulate debate rather than present a model of best practice. The resource is freely accessible via the NCETM portal[4]; interested readers may wish to view video clips illustrating the themes and activities outlined above. The site also hosts a 'toolkit' document commissioned by NCETM which guides departments through the resource and supports collaborative teacher-led CPD – concerning either pedagogical approaches or effective uses of technology. Our model of CPD is based on an iterative cycle of teacher-led discussion, peer lesson observation, collective reflection and refinement. Early trials in three schools point to the effectiveness of this approach in stimulating pedagogical reflection and change. (Bowker et al., 2009).

## Notes

[1] The T-MEDIA *(Exploring teacher mediation of subject learning with ICT: A multimedia approach)* project (2005–2007) was funded by the Economic and Social Research Council, grant ref: *RES-000–23-0825*.

[2] Organizational routines for managing use of the technology were also identified but space precludes detailing these here.

[3] This can be ordered in CD-ROM format at cost price via our website at www.educ.cam.ac.uk/research/projects/istl which also hosts further (existing and forthcoming) reports of this work.

[4] National Centre for Excellence in Mathematics Teaching: www.ncetm.org.uk/resources/7045.

## References


Bowker, A., Hennessy, S., Dawes, M. and Deaney, R. (2009). 'Supporting professional development for ICT use in mathematics using the T-MEDIA multimedia resource'. In M. Joubert (ed.), Proceedings of the British Society for Research into Learning Mathematics, vol. 29, (1) pp. 19–24. Cambridge.

Godwin, S. and Sutherland, R. (2004). 'Whole class technology for learning mathematics: The case of functions and graphs'. *Education, Communication & Information,* 4 (1), 131–52.

Goldenberg, E. P. (1995). 'Ruminations about dynamic imagery'. In R. Sutherland and J. Mason (eds) *Exploiting Mental Imagery with Computers in Mathematics Education,* pp. 202–24. Berlin, Heidelberg: Springer-Verlag.

Hennessy, S. and Deaney, R. (2007). *Teacher Mediation of Subject Learning with ICT: A Multimedia Approach (RES-000–23-0825).* Final Report to ESRC.

—(2009). '"Intermediate theory" building: Integrating multiple teacher and researcher perspectives through in-depth video analysis of pedagogic strategies'. *Teachers College Record,* 111 (7), 1753–95.

Hoyles, C. and Noss, R. (2003). 'What can digital technologies take from and bring to research in mathematics education?' In A. J. Bishop, M. A. Clements, C. Keitel, J. Kilpatrick and F. K. S. Leung (eds), *Second International Handbook of Mathematics Education,* pp. 323–49. Dordrect, The Netherlands: Kluwer Academic Publishers.

## Part Four

# Description of a Range of ICT Tools

Digital technologies, and their applications, are being developed at a great pace and throughout the world. Many of the 'drivers' for such developments have little or no obvious connection with education, at least at first sight. One large part of the market is for home entertainment, which includes gaming, simulation and the use of control devices – many now wireless – like paddles, bats, steering wheels, and so on. Another market is for mobile communications including telephones, cameras, internet, audio/video streaming, social networking, blogs, podcasts, and so on. A third, particularly for software development, is commercial use by employees in companies, organizations, R&D, and so on. Contributors to this chapter have been given a brief to highlight digital developments they see as having most to offer to education in general and to (teaching and) learning mathematics in particular.

Chapter 11

# Emerging Technologies for Learning and Teaching

Vanessa Pittard
*Director of e-Strategy at the British Educational and Communications Technology Agency, UK*

## Introduction

### Rapid Change

Looking back over the last ten years we can see a period of rapid technology-based change in learning and in society more generally. The internet and networked computers have led to significant shifts in how information and resources are accessed and developed. And for children and young people, their teachers and citizens more generally, personal access to computers and the internet is increasingly considered a norm.

In addition there is an increasingly rich range of devices and tools which can support learning in a variety of ways. This section looks at new and emerging technologies for learning, exploring the opportunities they present for what teachers can do to build more effective learning.

### Technology versus Practice

In any analysis of the role of technologies, however, it is important not to be beguiled by the technology itself. It is more important to remember that the first aim in adopting and developing any technology is that it is there for learning. So any consideration of future technologies should be framed to reflect that, taking into account not only how learning happens, but also the contexts of learning – schools, homes, classrooms and curricula. It also needs to set the benefits to learning against costs of implementation and change.

Becta has recently funded a major project led by the University of Nottingham and SERO Consulting to look at effective modes of learning that can be enabled by technology. In a recent report from the project (Sharples et al., 2009), the authors cite Richard Clark's point that technology ('media') will not change learning. Rather, it is change in practice that will do so (Clark, 1983, 1994).

Technology offers significant new *opportunities* to change practice. That is, develop and improve the interactions of learning between learners, teachers, symbols and objects. It is increasingly the role of educators and planners to understand technology-based opportunities for learning and translate these to possibilities and developments within their own contexts. By taking a considered approach to this, real benefits for learners can be achieved.

## New and Emerging Technologies

Becta's 'emerging technologies for learning' site[1] draws together news, research, analysis and views around technology developments relevant to education. It provides an environment for debate on technology futures within the education community and those serving it, encouraging dialogue and building shared understandings about the future. Further research studies on technology in practice are published on the research pages of the Becta website. The following sections draw on developments identified through these services, so thanks go to the various authors and Becta research teams who have provided contributions and analysis. I've presented these developments are presented in no particular order.

### Virtual Worlds and Simulations

The term 'virtual world' is a broad one, ranging from simple interactive micro-worlds, to more complex simulations and immersive environments. What they share in common is the power of computing and multimedia to present an analogue of a world that users can interact, manipulate and experiment with. Such systems are not new to maths. However, system capabilities are developing rapidly, driven by increasingly powerful computing environments leading to greater handling of complex data, rich multimedia, direct manipulation, environment construction, and sharing and collaboration.

These capabilities are particularly powerful when combined with common features such as the ability to overlay or present symbolic (e.g., mathematical) data and concepts alongside real phenomena. As Gorse et al. (2009) observe based on their study of virtual worlds tools for maths, such linking is especially important to help learners with real and applied maths.

Sarah deFreitas presents several examples of curriculum-based learning in virtual worlds which illustrate the potential of such environments (deFreitas, 2006, 2008). The use of rich virtual worlds is increasingly extending beyond professional training and development, with which they have been associated, to school and college learning. This also includes examples of effective learning in immersive social environments such as Second Life (Twining, 2007), which can support collaborative problem solving and higher order reflection.

## Games-based Learning

Games-based learning is closely related to virtual worlds and simulations. Again, the use of games to support maths learning is not new. Instructional games, for example, have been around for a long time. However, as with virtual worlds, game-related technology is developing rapidly. In addition to improvements in interactivity, multimedia and online collaborative tools, the last few years have seen learning become a greater focus in leisure-based gaming (for example, 'brain-training' approaches and virtual learning games). As a result, there are now several games-based learning projects in education using commercial games.

Paul Pivec (2009) has recently taken a critical look at games-based learning as a whole, including some of the claims made for the use of games-based approaches in education. Overall, he is positive about the potential of games in education, citing positive evidence of increased time on task, willingness to return to learning tasks, engagement of difficult to engage learners, and cognitive and higher-order learning benefits. Futurelab's recent report for educators on young people, games and learning (Williamson, 2009) is also positive, particularly in terms of strengthened learning relationships between teachers and students. Pivec's conclusion is that the real value from games-based learning comes from teacher-led, structured activities taking place around the games. This is 'games-based teaching'. This is true for new environments more generally, including virtual worlds and related developments such as augmented reality[2].

## Personal and Mobile Devices

One of the most striking trends of recent years is the development of personal mobile computing, whether through hand-held devices such as enhanced mobile phones or the use of netbooks and cut down laptops. Coupled with increasingly common mobile wireless access, we are moving closer to continual access to computing power and networks. David Ley (2007) points to the trend continuing beyond personal access towards pervasive computing, whereby technology becomes weaved into everyday life to an even greater degree. As technology develops, we will become more used to accessing networks and information on demand, and this will have an impact on how we work and learn.

Meanwhile the use of personal devices in classrooms in real time has delivered important opportunities for educators. Becta has supported research into the use of 1 to 1 hand-held devices for primary and secondary students (McFarlane et al., 2009). In both cases the students used the devices to support learning in a range of ways, including general internet search, taking photos, making presentations and accessing resources in class. Students were enthusiastic about personal access, using terms like help/helpful, fun/enjoy/enjoyable, interesting, easier, useful and handy. Access to the internet (not having to go to the computer room or compete for the computer at home) was a key benefit.

Elizabeth Hartnell-Young's study of the use of mobiles in school reinforces findings from the one–to-one hand-held evaluation (Hartnell-Young, 2008). Implementing personal mobile technologies for all needs to be led by the school and guided through appropriate policies, for example relating to internet access policy and acceptable behaviour. But as personal technologies develop, more schools will look to these to support learning.

## Display Technologies and Interfaces

Educational researchers such as Liz Burd and her team at Durham University are experimenting with one example of next generation of interactive display technology – Lumin tables[3]. These displays take the technology we are now familiar with in hand-held devices such as Apple's 'i-Pod touch' and make it larger, and responsive to multi-touch. The Durham project will work with schools to build software tools and develop school-based uses.

Michael Haller has taken a look at the range of related technologies, exploring their potential to improve interaction with computers and facilitate collaborative activities in more natural and intuitive ways (Haller, 2008).

He argues that this intuitive interactive quality is important for learning and that these interactive interfaces could help students to become more actively involved in working together with content and could also improve whole-class teaching activities. It is certain that these technologies have considerable appeal for learning and teaching, and potential to support both learners and teachers interacting with objects and with each other in different ways.

Haller offers a view on other promising developments in interfaces and display technologies such as interactive pens and paper. These technologies support of information exchange between computer and non-computer devices (e.g., enabling quick digitization of what students write in exams). Though it is unclear when or how these technologies might come into use, they may have potentially useful applications in education settings.

### Online and Blended Learning

As with other developments listed above, online learning is not a new concept. However, changes to the context of learning mean that there is a new climate for online and blended learning. For example, most young people now have good personal access to the internet and are comfortable leading many aspects of their lives online. The 'science' of online learning has developed, and this includes models of tutor/mentor support and building learning communities. Though not all students will benefit, online learning can support 'catch-up' and improve differentiation and choice in learning.

Becta recently looked at emerging uses of online and blended learning by schools, commissioning Manchester Metropolitan University to develop models of effective use based on evidence (Lewin et al., 2008). In a context where individual tuition is becoming increasingly important, especially for learners who are falling behind, online solutions may in fact provide a way of freeing some teacher time to focus on just that. Schools will need to make decisions locally. However, it is clear that online learning will become an increasing feature for school-aged learners, whether used formally or informally. Teachers will need to make judgements about what will be truly effective in supporting learning.

## Conclusion

There are a range of ongoing and new developments in learning technologies, many with potential to enhance and transform maths teaching.

Technology tools provide the opportunity to deliver truly excellent learning and teaching as never before. Some of these, as with multi-touch surfaces, are relatively new and their development needs to be tracked and understood. Becta will continue to share evidence and support discussion by teachers about the potential and role of these developments for learning.

What is clear from more developed approaches, like serious games and online learning is, as a 2009 Australian Learning and Teaching Council report concludes, learning technologies need to be clearly integrated with curriculum and assessment to add real value to learning. It is the role of schools and teachers to discuss how that can be achieved and plan well for a future in which technology is part of the fabric of learning.

## Notes

[1] http://emergingtechnologies.becta.org.uk/

[2] http://emergingtechnologies.becta.org.uk/index.php?section=etn&rid=14696

[3] http://tel.dur.ac.uk/?p=23

## References

*Australian Learning and Teaching Council (2009).* 'Educating the net generation: A handbook of findings for practice and policy' www.netgen.unimelb.edu.au/outcomes/handbook.html (accessed on 5 February 2011).

Clark, R. E. (1983) 'Reconsidering research on learning from media', *Review of Educational Research* 53 (4), 445–459

—(1994). 'Media will never influence learning'. *Educational Technology Research and Development,* 42 (2), 21–29.

deFreitas, S. (2008). 'Emerging trends in serious games and virtual words'. SGI/Becta http://emergingtechnologies.becta.org.uk/index.php?section=etr&marker=3&filter=ArtTec_001

—(2006). 'Learning in Immersive worlds'. JISC. www.jisc.ac.uk/media/documents/programmes/elearninginnovation/gamingreport_v3.pdf

Gorse, C., Dickinson, I. Shepherd T, Whitehead, P and Walker, C. (2009) 'Virtual maths in an industrial context: Learning with new and emerging technologies, Final report', BECTA. March 2009

Haller, M. (2008). 'Interactive displays and next generation interfaces'. Becta. http://emergingtechnologies.becta.org.uk/index.php?section=etr&catcode=ETRE_0001&rid=14164

Hartnell-Young, E. (2008). 'How mobile phones help learning in secondary schools'. Bectahttp://partners.becta.org.uk/index.php?catcode=_re_rp_02&rid=15482&section=rh

Lewin, C., Whitton, N., Cummings, J., Roberts, B., Saxon, D., Somekh, B. and Lockwood, B. (2008). 'MILO: Models of innovative learning online'. Becta. http://partners.becta.org.uk/index.php?section=rh&catcode=_re_vi_hi_03&rid=15545

Ley, D. (2007). 'Ubiquitous computing'. Becta. http://partners.becta.org.uk/upload-dir/downloads/page_documents/research/emerging_technologies07_chapter6.pdf

McFarlane, A, Triggs, P and Ching Yee, W (2009) Researching Mobile Learning: Overview, Becta.

Pivec. (2009). 'Game-based learning or game-based teaching?' Becta. http://emergingtechnologies.becta.org.uk/index.php?section=etr&rid=14692 (accessed on 5 February 2011).

Sharples, M., Crook, C., Jones, I., Kay, D., Chowcat, I., Balmer, K. And Stokes, E. (2009). New modes of technology enhanced learning: opportunities and challenges. Becta.

Twining, P. (2007). 'The schome-NAGTY teen second life pilot'. Final Report. Open University. http://kn.open.ac.uk/public/getfile.cfm?documentfileid=11344

Williamson, B. (2009). 'Computer games, schools, and young people: A report for educators on using games for learning'. Futurelab www.futurelab.org.uk/resources/documents/project_reports/becta/Games_and_Learning_educators_report.pdf (accessed on 5 February 2011).

Chapter 12

# Home and School – Bridging the Gap

Russell Prue
*Educational Consultant and Journalist*

## Introduction

Before we start to look at tools for our trade as mathematicians, we need to briefly examine how Information and Communications Technology (ICT) is being used within our organization and establish a strategic way forward. Pockets of innovation are hard to maintain without a holistic approach to your organization's use of learning and teaching technology. It is for this reason that I am not going to be covering the latest developments in virtual on-screen geometric drawing tools, and neither will I comment on which spreadsheet is best for maths. It is more important to consider how these and other tools are used within the context of learning and teaching.

Everything we know about everything is available online. It is possible to teach a child a full and varied curriculum without ever going near a school. The internet has demographically shifted the access to learning and Web 2.0 tools, *Facebook, Youtube, Twitter* continue to shape the way information is presented. Traditionally we would go to a library, use their indexing system to search on keywords, subject categories, titles and genres. Using the Internet and Web 2.0 tools, every word is now a keyword, every article can be searched and linked in ways never before considered. Knowledge is cheap. Creativity and problem solving skills are the new measures of success. It is not enough to say that our use of ICT is exemplary just because we have replaced our dry marker boards with an interactive whiteboard, many of which are simply not being used as effectively as intended.

While we continue to insist on measuring memory and not the true ability of the learner, the real potential of ICT will continue to evade us. Perhaps that might continue to happen for many years to come but I doubt it. Early indications are that new exams are emerging and with OCR leading the way with modular, online, paperless tests that are taken when the learner is ready there is still hope. In lots of classrooms right now many of our young learners are simply missing out on the opportunity of using the true potential that ICT offers learning and teaching. We know that when technology is left in the hands of learners it is often used in exciting and creative ways that many educators do not appreciate. Quite different to the methods and adoption classroom practitioners have been encouraged and forced to follow. Often technology is simply used to replace older methods of working. Maths text books have been replaced by online worksheets and sometimes by crude software.

If, for just one moment, you accept that we are lost in a sea of technology and are asking someone for directions to the promised land would we be surprised to hear 'you don't want to start from here'. Perhaps it's just not possible to start from anywhere else and the best that we can do is while on this journey is to ask some really difficult and unpopular questions about how and why we are using ICT. To successfully embrace some of this new technology in our learning and teaching we must be prepared to stop doing what we have always been doing. Often that means stop trying to bolt stuff on and instead take a radical look at what we are trying to achieve.

Mathematics has had, and will always have, an inseparable link with computing, after all everything we do with digital technology has been developed from an understanding of maths. Take away FREE Mobile Phone GPS tracking within *Google Earth*, take away social networking and everything else we do with computer based tools and all you have left are a series of mathematical calculations based on the crudest of all languages, Binary!

I think that while educators need currency to buy learners' attention; (leaving aside the moral and political dimension to this), there is a valid case for using technology. This can and often does have an immediate and motivational impact. The concept of social networking is incredibly popular with young people. One very interesting tool is *Twitter*. It is a micro blogging tool, it is free to use and interestingly it is no longer being used by only young people. There are lots of explanations for this. I think that adults got to *Twitter* early, unlike '*MySpace*' – which should really be called '*TheirSpace*'. It just isn't cool to hang out in the same place as adults. *Twitter* is a community of bloggers who, instead of writing for a webpage, write a short (up to 140 characters) description of what they are doing. This is then posted onto a public time-line and by choosing whose tweet you see, you determine the type of information that is pushed towards you. There are other aspects to this brilliant C21st communication tool and you can be excused for perhaps overlooking them. The British media would have us all believe that *Twitter* is a waste of time and that you have to be an absolute 'Geek' to want to use it. While I am sure that there are some outstandingly Geek-ish aspects to *Twitter* as there will be with any social network, and at this very minute there is someone commenting on the thickness of the butter being applied to their piece of toast, these are not good enough reasons to dismiss *Twitter*.

How can you use *Twitter* in the classroom? Simply by asking your followers a question you can get real instant answers that you can use in your teaching. Any subject could benefit but let's stick to maths; I know that some of the early adopters of *Twitter* have used it to bring real-life statistics into class. Simply ~~by~~ asking 'how likely is rain today where you are?', generates a flurry of intense Tweeting activity and loads of useful data. Having location information available too allows learners to plot this information in *Google Earth* or on a traditional map. The activity is real, relevant and reflective and loads of fun to do. The educator is then left to facilitate the information retrieval and ensure the learning objectives are met. This is disruptive technology and if you weren't ready for it, this activity will turn your classroom upside down, never a good enough reason not to do it even when the Inspector calls! It isn't going to work if your establishment bans *Twitter*. The activity can be applied to lots of mathematical activities such as solving puzzles, getting help from experts, and collaborating with others in data rich projects.

Our Learning Platforms and Managed Learning Environments also have the capacity to improve learning outcomes but only if they are used appropriately and not simply as a worksheet information store. John Davitt often illustrates this point with a brilliant analogy between education and shopping. Why is it more acceptable that online shopping environments are more developed than our online learning environments? When did you last hear of a computer system offering the learner extension activities based on their current level of interest? 'Learners that found the Right Angled Triangle interesting also found the Isosceles Triangle Interesting?' Until this is common we won't have properly addressed the use of ICT in learning and teaching.

I like technologies, or uses of technology, which challenge established practice. It is not good enough that we continue to carry on doing what we've always done without asking is this the best way for the learner to proceed. Collaboration is one of the most important survival skills for C21st and it is a great shame that we don't assess this skill or celebrate it in our assessment of learners. Imagine turning up for a public exam with three friends, a Net-book and a mobile phone and being welcomed in!

In combining some pretty traditional learning software for arithmetic, for example, www.sums.co.uk, and a more radical use of your whiteboard you can enhance the learning outcomes otherwise associated with them separately. By turning your whiteboard through 90°, laying it flat facing up on top of a table and by using a short-throw projector you have an interactive touch table surface that many learners can gather around without standing in front of the projected image. By handing some of your technology over to the learners you shift the balance of power within the learning space and what you will see happen is just short of miraculous. Learners will talk, share and take it in turns to touch the board. They will argue, defend and explore new concepts and ideas but most importantly they will be collaborating and learning together. You can buy some of this technology ready-made from Microsoft and also from Smart Technology or from Matrix Displays with a 'Vipro' glass table, one of the biggest touch surfaces available providing a 67-inch diagonal image.

Taking the class to the 'ICT Suite' can often be a traumatic experience, for everyone! Whole class instruction using a room full of 30 desktop personal computers was never very satisfactory, except perhaps in the very early days of ICT Suites. Now, learners have different and varied levels of expertise and it is almost impossible to instruct at a consistent level, without some of the learners rushing off to explore different areas of the software you are trying to use as a group. Instead the technology needs to come to you, to your space and it can do that in the form of very inexpensive UMPCs (Ultra Mobile Personal Computers). There are lots of brands to choose from with some highly recommended ones starting at just £99. In my school of the future, you wouldn't be responsible for these either; I think laptop trolleys have done more harm than good. It is now clear that your learners should be able to produce UMPCs fully charged, ready to go and at any time they are needed. Establishments that have done this by equipping young people with this technology, and which have also empowered them to take control of their maintenance, are benefiting from a truly transformational use of ICT in all subjects. In real terms this means mains power available in learners' lockers for recharging so this isn't something that you can just bolt-on quickly for the maths department, but it is something to aim for as a whole organization.

## Conclusion

Greater levels of transformation will occur when mathematics is merged with other essential key subjects like science and engineering. Elsewhere in this book you will find references to the current STEM (science, technology, engineering and mathematics) strategy. This is an initiative designed to encourage greater take up of each of these subjects, ideally through applying them together in realistic contexts. Obvious when you think about it, really. This leads nicely onto data logging and some very interesting developments by Texas Instruments who continue to push the boundaries of classroom calculator innovation, with some of their most recent products looking more like mini-networking devices that allow content and material to be shared and exchanged between learners. A data logging USB probe can be easily attached and sensor data displayed and manipulated on the same device. Educators can also ensure compliance with exam acceptance as the TI-Nspire comes with a functionality reduction feature for use in exam situations. The Teacher edition software also runs on a PC and has Q&A functionality built in.

However you choose to use ICT in your learning and teaching spaces you will find that by empowering learners to take control of ICT you will free yourself up to concentrate on the importance of really good maths teaching.

Chapter 13

# Personal Portable Technology

Adrian Oldknow
*Emeritus Professor, University of Chichester, UK*
Peter Hamilton
*Head of Educational Development, Intel IT Innovation, Ireland, UK*

## Introduction

There's nothing new about people wanting to have technological devices that they can take around with them. The compass, pocket watch, slide rule, sextant, telescope, camera, gramophone, transistor radio, cassette recorder and pocket calculator were all artefacts which at their time made great impact on society – many of which still survive. Now we are used to a compact battery-driven array of devices which include mobile phone, digital organizer, digital still/video camera, MP3 music player, GPS navigation, internet browser, e-mailer, radio/TV receiver, alarm clock, game station, health monitor and laptop computer – often with several functions combined in the same unit. But their current impact on mathematics education, at least in the United Kingdom, is still more or less negligible. This is about to change radically, at least in many developing countries and also in most curriculum subjects, thanks to recent developments in technology. Now as we look forward the rapid advance of Moore's law continues unabated which is doubling the power of computer microprocessors every 18 months while also reducing their price and power consumption is providing an increase in capacity and mobility never before seen in human technological development. It is this change, mirrored in all components of the modern mobile computer including storage capacity and screen performance and cost, that is driving the constantly reducing price and increasing portability of these devices. Even more rapidly, the available telecommunications bandwidth is increasing at double that pace with capacity at a given price point doubling every 9 months due to advances in the similar underlying physics of optical circuits and electronic switching technologies. Mathematicians will understand the impact of these two exponential functions much more than the general public so we have no excuse for being unaware of this great potential and the opportunity it presents for our economies and societies. Before discussing the potential impact of the new wave of personal portable technology we have a short review of what has come before it.

## A Short Review of the Past Three Decades

We have had cheap and portable, so-called four-function, and scientific, calculators widely available since the early 1980s, and yet it took nearly 20 years for them to be integrated into the English National Curriculum. We have had graphical calculators available since 1986 (the Casio fx-7000g) which at first made some impact on post-16 mathematics teaching – with the MEI, SMP and Nuffield A-level schemes all integrating their use into course materials. However, in a period of mathematical neo-Luddism (2001–2006), the assessment scheme at AS/A2 was changed to ban the use of graphical calculators in about half the examinations. Although this is no longer case, and students may now use them in all examinations, it does not appear that students currently receive the encouragement to use graphical calculators is they did in the 1990s. We have also had symbolic manipulation (CAS = Computer Algebra Systems) available in hand-held devices since the mid 1980s (the HP-28g), but while many other countries have been evaluating their use as a learning aid in mathematics, the English authorities have fought shy of grasping that nettle.

In 1993 the National Council for Educational Technology (NCET – precursor to Becta, the British Educational Communications and Technology Agency) set up an ambitious Portable Computers in School Project with projects involving primary and secondary school pupils in nearly every LEA and curriculum area. So, for example, Hampshire LEA had a mathematics project in three 11–16 secondary schools evaluating the impact of three different forms of technology. One school had sufficient TI-81 and TI-85 graphical calculators for two parallel classes to use in pairs at the same time. The TI-81 was, for its time, a very powerful device – but it lacked any form of I/O (input and output) and so could not be projected, or plugged into a computer or printer. The TI-85 was more expensive but could be connected to an LCD display panel for whole class projection, and could exchange data between units, as well as with computers and printers. At that time compatible data loggers were not yet available. Another school had a set of 15 laptop computers, which were then heavy, very expensive, had a relatively short battery life between recharges and had a weak spot where the power cable was plugged in and out. These could be used for whole class display with a large colour monitor. The third school had a set of 30 HP-95 palm-top computers with built-in graphing software, a spreadsheet and the Derive mathematical software on a memory card. Thus a whole class was equipped with these for a year, and students were able to take them to all lessons and keep them at home after hours and during vacations. There was no means of whole class display, but there was a serial cable for connection to PC and/or printer. Students quickly started using the calendar and word-processing functions for keeping homework diaries and writing up notes from other lessons. It is interesting to see in many ways how little things had moved on in term of the integration of portable Information and Communications Technology (ICT) in schools in the 10 years from then until the 2003 Becta research report (Becta 2003).

The DfES Maths Alive! research project (2000–2001), reported on by Don Passey in Chapter 6, deliberately went for a 'mixed economy' where the teachers at 20 pilot sites had laptops and printers, the classroom had 3 or 4 desktop PCS and a ceiling mounted projector together with an analogue interactive whiteboard, and there was also a set of TI-83 graphical calculators for use in pairs on students' desks. The teachers also had TI-83s for projection with the LCD pad as well as a data logger: a TI-CBR for collecting distance-time data. Thus teachers had considerable flexibility in the kinds of ICT tools they could deploy – but, perhaps not surprisingly, most preferred to work in a whole class, teacher-led fashion (Passey 2005).

Professional associations such as the Association of Teachers of Mathematics and the Mathematical Association, and groups such as Teachers Teaching with Technology (T3), have provided support and encouragement for mathematics teachers to provide hands on access to ICT including the use of graphical calculators, and video case studies produced by the DfES (now DCSF) illustrated their use in classrooms. While some of these materials are now getting rather dated – at least in terms of the fashions worn in the photographs – their content remains just as relevant now. See, for example, the Becta book on data logging (Oldknow and Taylor 1998). A more up-to-date source is the learners' entitlement document (Becta 2009). In general, though, there has been little incentive for mathematics teachers to provide hands on access to ICT – except, that is, until the compulsory ICT-based coursework task for data-handling in GCSE mathematics was abolished. This is in stark contrast to the way personal technology has been integrated into the International Baccalaureate.

## Tentative Steps

For a while it appeared that a technology known as the PDA (Personal Data Assistant) might hold the key to getting hand-held learning widely adopted. Wolverhampton LEA piloted a project using PDAs in four primary schools in 2003. Parents bought the devices for their children under a lease-lend scheme – and so the pupils could use them at school and at home. While they had a limited number of educational applications available, the PDAs were used extensively as web-browsers, as well as cameras, and pupils were able to produce project presentations: www.learning2go.org/.

Internationally Nicholas Negroponte from the Massachusetts Institute of Technology launched a $100 laptop computer, The Children's Machine, in November 2005, designed for students in the developing world. This provided a challenge to the ICT industry to produce high-specification, ultra-portable but rugged devices suitable for education worldwide. An early example of this was Intel's first Classmate PC from 2006 which was piloted in South America and Southern Africa as part of their World Ahead Program.

A different approach has been taken in the development of the TI-Nspire hand-held device. Closer in price to a graphical calculator, the TI-Nspire is not designed as a web-browser, but has its own integrated mathematical equivalent of Office software, in which a document can consist of several pages which might contain calculations, graphs and geometry, lists and spreadsheets, data and statistics, notes, data-capture from sensors, and so on. Files can be exchanged between units, and with computers. In the classroom several units can be connected to the computer at the same time, and there is also a new classroom wireless network system, TI-Navigator. There is a software emulator for the hand-held which runs on Windows and Mac computers and which can therefore be used for whole class displays. A pilot project using TI-Nspire in seven secondary schools in England was completed in 2008. The project evaluation report is available (Clark-Wilson 2008). Activities from the project are available too (Clark-Wilson and Oldknow 2008).

Increasingly UK schools are issuing Windows laptops to students, and we are fortunate that in mathematics there is already a very powerful and appropriate software base to support personal learning. Many schools used the opportunity provided by the so-called 'e-Learning credits' to purchase site-licences of software such as *Autograph, Geometer's Sketchpad, Cabri,* and so on, even if they have not been used extensively to date. Most laptops have a spreadsheet installed – either MS *Excel* or *Open Office,* and there is a range of powerful free, or very cheap, software to support mathematics, such as *Geogebra,*the MA/Intel Skoool *Mathematical Toolkit, Tracker* and Vernier's *Logger Pro.*

## The Tipping Point

At the Hand-held Learning conference in London in October 2007, the UK educational ICT company, Research Machines plc, announced the first Asus minibook PC for under £200. Within the last 2 years we have seen a plethora of ever more powerful such ultra-portable notebook PCs arrive on the market with prices between £200 and £300, about a tenth the cost of the first laptop computers! Most of these are now preloaded with a version of Windows, normally XP.

In January 2009, Intel announced their third generation of the Classmate PC, as a tablet PC with a touch sensitive screen. In the United Kingdom it is sold through a company called Zoostorm, and the current top of the range model, called the Fizzbook Spin 10.1" PC, is powered by a new Intel Atom processor: www.besa.org.uk/besa/news/view.jsp?item=2197

Like the Apple I-phone it has accelerometers inside which detect whether it is being held in landscape or portrait mode. It is also equipped with an integrated camera and wireless internet connection as well as a 60Mb hard drive and Windows XP. At around £300 ex-VAT it is currently being evaluated by members of the Association for Science Education (ASE) and the Mathematical Association (MA) as a baseline platform to support secondary school (11–18) maths, science and STEM teaching and learning. In addition to the mathematical software already mentioned (e.g., *TI-Nspire, Autograph, Sketchpad, Cabri, Geogebra, Mathematical Toolkit*), there are statistical tools (*Fathom*), modelling tools for science (*Modellus, Tracker*), physics simulations (*Algodoo*), data logging tools (*Logger Pro, Coach*) and programming environments (*Alice 2.2*). Used in conjunction with USB and wireless sensors (Vernier, TI), as well as high speed cameras (Casio Exilim EX-FH20), we have an undoubted Rolls-Royce of the ICT world at the price of a Model T Ford! Intel is now working with a group of UK and Irish maths and science educators to develop other forms of simulations and interactive environments to support innovative teaching and learning approaches, which include support for robotics (Lego *Mindstorms*) and digital prototyping (*Arduino*).

## It's a digital world for students – let's get it joined up

Anyone who has come into contact with youngsters, say 3–7 years old, cannot help but notice the speed with which they pick up the use of technology – particularly that associated with video games and watching TV. Older students already make extensive use of technology such as the web and mobile phone for texting, Skyping, Twittering, blogging, podcasting, sharing photos and videos, social networking and other such absorbing pastimes. They can certainly be forgiven for not making connections between any of these activities and the aspects of maths, science, technology and engineering which make them possible. Similarly, given the bewildering pace of technological innovation, it is not surprising that most of those involved in teaching them are also unaware of these connections.

It could be argued that most secondary school mathematics departments are now equipped with sufficient ICT tools (laptop, printers, internet access, data projectors, IWBs, learning platform) to support their teaching needs with the current curriculum. The floodgates opening up with latest educational ICT advantages give us the opportunity to concentrate on fulfilling the needs of the learner. Of course we can provide them worksheets to complete, or videos to watch, just as we have done for many years, but the real challenge is to find ways to use the ICT to excite their interest and involvement in learning – and awaken their curiosity in mathematics – as well as science and STEM. The trouble is that we don't have experience in doing this! But perhaps that is all for the better if we want to engage learners more actively in taking shared responsibility.

We already have some good examples of what to do with digital images. For example, photo and video editors provide their own mathematical environment within which transformations such as reflections become 'flip vertically' and 'flip horizontally'. You can rotate and enlarge images, and interpret a variety of measurements, such as pixels, mm, frames per second, in making things fit. Digital images, such as Richard Phillips' 'Problem Pictures' can be projected to prompt mathematical discussion – but students' (and teachers') own images can also be used for that purpose, such as bringing a lesson on symmetry to life. Digital images can be introduced as the background for graphing and/or geometric construction in software including *Autograph, Cabri II Plus, MS Excel, Geogebra, Geometer's Sketchpad, Mathematical Toolkit*. Video clips can similarly be used as stimuli. Displayed on an IWB, points can be 'captured' as the clip is single-stepped, and conjectures made about the path of a moving object. With software such as *Coach, Logger Pro, Mathematical Toolkit* and *Tracker* students can capture and model data from videos of their own activities, such as throwing a netball, or bouncing on a trampoline. The new range of digital cameras with high speed video mode allows movements to be captured which could not be seen with the naked eye.

Animations, cartoons and computer games use graphical techniques for virtual reality based on computer graphics (applied geometry) techniques originally developed for computer aided design (CAD) – such as Bézier curves – and many of these are quite accessible applications of school geometry, algebra and calculus. They also use so-called 'physics engines' so that things appear to move according to physical laws. (Just look at the amazing love scene in space in Pixar's WALL-E cartoon where one figure uses squirts of a fire extinguisher as his power pack.) It's just as important for students to be aware of the entertainment applications of maths and science as it is for the obvious disciplines of engineering, rocket science, and so on. These are certainly major sources of employment for mathematics and physics graduates – so even work can be fun (or *Phun* qv).

Mobile phones, Sky TV, GPS and the internet all rely on satellites in geostationary orbits launched by rockets such as the European Ariane. Sky TV is received by a parabolic dish which focuses rays through a collector in the same way as an optical telescope does. Journeys on foot, bike or car can recorded on GPS devices such as Garmin and the resulting data retrieved for analysis, for example, in Excel and Logger Pro, but also directly into Google Earth and Google maps, showing data such as speed, time, direction and altitude. GPS and other data logging devices, such as accelerometers, heart monitors, temperature gauges, radar guns, and so on can be used to record students' athletic exertions such as long-jumping, javelin throwing, running – and correlated with, for example, video data to model motion. The Nintendo-Wii games console already uses wireless data logging together with maths, physics and biomechanics to record a player's motions and to predict its effect on an abject, such as rolling a ten-pin ball, hitting a golf shot or kicking at goal. The Microsoft system uses a pair of stereo cameras as an alternative. We can expect to see many applications of this kind of technology in the future.

If we can come up with interesting and challenging tasks and ideas for students to pursue, alone or in groups, we can also provide them with opportunities to practice and improve their communication skills in conveying their findings.

## Conclusion

On 19 April 1965 *Electronics Magazine* published a paper by Gordon Moore in which he made a prediction about the semiconductor industry that has become the stuff of legend. Known as Moore's Law, his prediction has enabled widespread proliferation of technology worldwide, and today has become shorthand for rapid technological change. Today Moore's law continues to apply to the communications and computing industry where

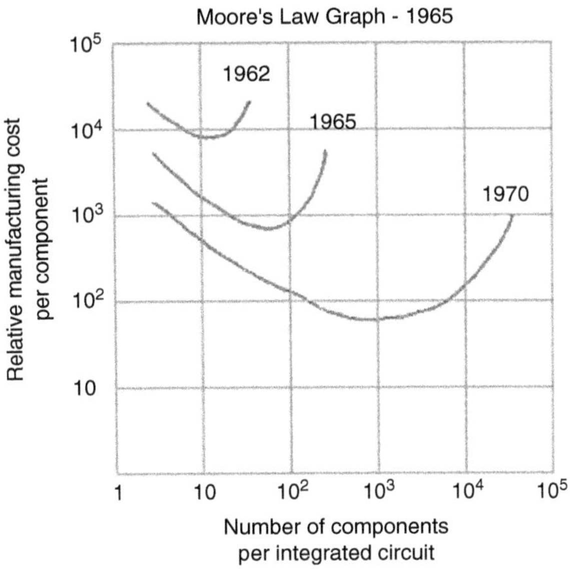


computing and memory capacity doubles every 18 months. Any fan of digital photography will have seen an almost 100-fold increase in memory card capacity at a given price point over the past 9 years from 64Mb to now 4Gb and above as standard. The increase in internet bandwidth driven by the same underlying microprocessor and optical physics is actually increasing at a rate that is twice as fast as Moore's Law, with bandwidth at a price point doubling every 9 months. We have seen connection speeds increase from 56 kbps dial-up less than 10 years ago to speeds for 40 mbps projected for WIMAX connections and 4G mobile connections over the next 12 to 18 months. The implications of this rapidly increasing capacity are far reaching with computing technology moving from being scarce to becoming abundant resources which are 'too cheap to meter'. We are only part way through this dramatic and fundamental transformation in widespread access to computing power.

Such developments in technology are producing an increase in availability and access to devices, content and communications at growth rates never previously seen in any industry or human activity in history. The continued advances in physics underlying the rapid advancement in semiconductor and communications technologies, together with the explosion of open content and software, are driving fundamental social and economic changes (Anderson 2009). It is critically important that our young citizens and members of the future workforce are using these technologies to learn, problem solve, innovate and take a leadership position in the modern smart economy – see (Oldknow 2009). Much will depend on whether we have the skills, and will, to maximize this opportunity for advances in education. How much do we risk if we fail to do so?

## References

Anderson, C. (2009). *Free: The Future of a Radical Price.* London: Random House.

Becta. (2003). *What the Research Says about Portable ICT Devices in Teaching and Learning.* Becta: www.education.gov.uk/publications/standard/publicationDetail/Page1/15002 (accessed on 5 February 2011).

—(2009). *Secondary Mathematics with ICT: A Pupil's Entitlement to ICT in Secondary Mathematics.* Becta: www.teachernet.gov.uk/_doc/13781/entitlement_sec_maths .doc (accessed on 5 February 2011).

Clark-Wilson, A. (2008). *Evaluating TI-Nspire in Secondary Mathematics Classrooms.* University of Chichester: www.chi.ac.uk/teachered/documents/Clark-Wilson 2008TI-NspireFinalReportv5.pdf (accessed on 5 February 2011).

—and Oldknow, A. (2008). *Inspiring Maths in the Classroom.* Texas Instruments: http://education.ti.com/sites/UK/downloads/pdf/Inspiring_Maths_in_the_classroom.pdf (accessed on 5 February 2011).

Moore, G. (1965). *Moore's Law.* Intel: www.intel.com/technology/mooreslaw/.

Oldknow, A. (2009). *Their world, Our World – Bridging the Divide.* Teaching Mathematics and Its Applications V28 N4, 180–95 December.

Oldknow, A. and Taylor, R. (1998). *Data Capture and Modelling in Mathematics and Science.* www.adrianoldknow.org.uk/datacapture.pdf (accessed on 5 February 2011).

Passey, D. (2005). *Maestro: MathsAlive Evaluation Study.* Lancaster University: www.rm.com/_RMVirtual/Media/Downloads/Final_Maestro_Project_Paper.pdf(accessed on 5 February 2011).

# Part Five

# Practical Ideas of ICT to Enhance Teaching and Learning

In this section, a selection of writers, who are engaged with teachers in various capacities, outline some of the Information and Communications Technology (ICT) rich resources that are available to enhance classroom practice. Many of the resources referred to are freely available. A small number require a subscription fee, but these are included as they are representative of a wider set of websites and resources and are available in many schools.

Chapter 14

# Linking the Mathematics Curriculum to Software, Resources and Strategies to Engage Teachers and Learners in Using ICT in the Classroom

Linda Tetlow
*Consultant*

## Introduction

In the autumn of 2003, after a 2-year break from mathematics teaching, I took on a new role as mathematics coordinator for a new and innovative project for online 'Out of school learning'. I have always been interested in using ICT to enhance mathematics teaching and learning and looked forward to both the opportunities and the challenge that teaching in an interactive online classroom would present. As coordinator I was required initially to prepare a programme of study for the students whose ages, mathematical background and reasons for being educated out of school varied considerably. Then I needed to prepare and upload my own lessons and materials, together with some prepared lesson plans and materials that could be used by other teachers. There were two principal challenges:

- Preparing my own materials that could be used in the on-line classroom using the facilities available and that would engage a range of learners, some quite reluctant. This might be using a very simplified interactive whiteboard which had the facility to grab snapshots of documents and web pages, using uploaded PowerPoint presentations and application sharing some software such as spreadsheets, graphing packages or dynamic geometry.
- Finding interesting resources that were available on the internet that we could share in the classroom to save time-consuming preparation and that could be used by the other teachers.

Despite trying to maximize the use of the available internet resources, because constantly preparing and uploading my own materials was very time consuming, I really struggled to find a good range of interesting and motivating material and certain mathematical topics were especially problematic. For example, anything 3-dimensional was difficult for students to interpret on a 2D whiteboard. I was working very much in isolation mathematically, although sharing general strategies with coordinators of other subjects, and would have welcomed opportunities then to share ideas with other mathematics teachers.

In the years that have followed I have had several opportunities to work with a variety of mathematics educators who share the belief that the use of digital technologies is an important part of teaching and learning mathematics and who have come together to pool their knowledge of where to find resources and how to use them in the classroom. There are lots of ideas, software and resources that I now know about and could use in a 'real' or 'virtual' classroom. The result of this more recent work has led me to two conclusions.

- The world has moved on since 2003–2004 and there are now far more resources available to mathematics teachers both software and internet resources. Additionally the increased speed and widespread availability of internet access has opened up enormous possibilities for teachers, learners and educators to use and share this software and resources.
- As a teacher working in isolation mathematically I could not possibly have discovered all that was available. Some means of consolidating shared information, linking it to aspects of the mathematics curriculum and making this information available for all teachers to access easily is essential.

I had the opportunity to do this when I helped to coordinate the work a group of mathematics educators who attended a 2-day brainstorming session. The purpose of this session was to share ideas about useful ICT resources and classroom tested ideas and strategies and to link these to particular areas of the mathematics curriculum with an emphasis on 14–16 year olds. The *'ICT enriched curriculum grid'* which links the curriculum to activities and resources involving the use of ICT is available by joining the National Centre for Excellence in Teaching Mathematics (NCETM) website. Once you have joined and logged in, select 'community', then 'view all communities', and then 'ICT in Mathematics'. You need to apply to join the community but any logged in users of the site can do this. Finally select the documents tab and 'BECTa ICT products'.

The new mathematics curriculum in England (2007) lists under 'Curriculum opportunities' that pupils should 'Become familiar with a range of resources, including ICT, so that they can select appropriately' and expands on this to say that 'This includes using practical resources and ICT, such as spreadsheets, dynamic geometry, graphing software and calculators, to develop mathematical ideas'.

I have included in what follows a selection of ideas which I hope show the potential of ICT to enrich the curriculum and to provide opportunities for work which would be very difficult without it. I have tried to include examples which include a wide range of tools and resources and include a variety of areas of the mathematics curriculum. Wherever possible the examples shown are linked to further information which includes possible lesson materials and classroom resources, help with using software and in some case to actual examples of classroom use by both experienced and less experienced practitioners.

## Objective: Work on Tasks That Bring Together Different Aspects of Concepts, Processes and Mathematical Content

In this case:

- Number and Algebra – Graphing straight lines and gradients
- Geometry – Properties of 2D shapes; Points, lines and shapes in 2D coordinate systems

### Match the Graph

In this activity students use a graphing package or graphical calculators to input equations of straight line graphs to match a given image. For example, to form the shape of a kite or a parallelogram:

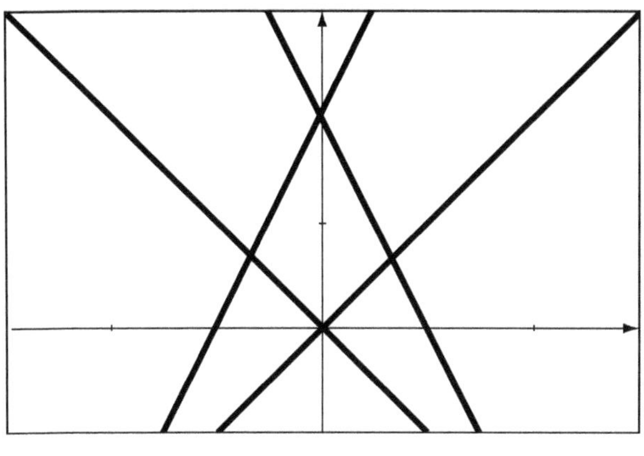

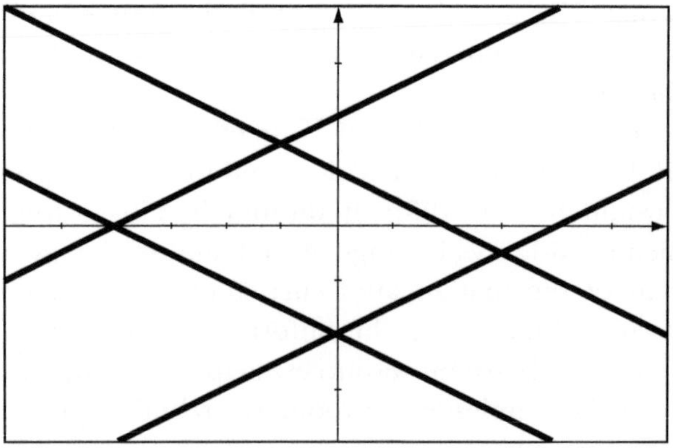

Using the graphical calculator or graphing software enables the pupils to generate and explore graphs of linear functions quickly thus enabling them to focus on the key properties of the functions and how they can be used to create geometrical shapes with particular properties.

This task is 1 of a set of 28 activities in the '*Practical Support Pack*' which includes lesson plans, classroom resources and practical advice and was developed with the support of The Mathematical Association. They cover a wide range of mathematical content and provide opportunities for pupils to develop their problem-solving skills, mathematical strategies and thinking skills.

## Objective: Visualize and Work with Dynamic Images

### Visualizing in 3 Dimensions

This can be very difficult without the aid of models, at least initially. It is easy to underestimate the difficulties of interpreting 2-dimensional representations of 3D objects and the misconceptions that can arise from these. 3-dimensional dynamic geometry software such as the powerful *Cabri 3D*, or the simpler *Yenka 3D shapes,* offer the facility to construct solids quickly and to open them to reveal their nets or to rotate the viewing angle using the mouse to view them from different directions. These are some examples:

- Visualizing right-angled triangles in order to use Pythagoras' theorem or trigonometry in 3D
- Visualizing plans and elevations of a 3D object

- Constructing nets of common solids and visualizing which edges or corners connect

An example of *Cabri 3D* being used in the classroom is shown in the Teacher's TV programme 'Hard to teach – secondary Maths using ICT'. The lesson shows a classroom where pupils are trying to 'Find the length of the longest stick that will fit into a cuboid'. In this video pupils are using the software on an interactive whiteboard linked to a single computer. There is more information about this lesson activity in the 'Inspire me' section of the BECTa website. This lesson can also be seen, together with further information, activities and links to on-line tutorials on the Cabri website. A different classroom set up and activity is shown in an overview video which can also be viewed on the site.

There are also some very useful applets available on the internet for specific 3D applications. **The Standards unit pack** *'Improving Learning in Mathematics'* is a multi-media resource which has activities that encourage learners to become more independent and reflective about their mathematics. The pack includes an activity *Building houses* **which allows students to explore** connections between 3-dimensional models and their plans and elevations, using the mouse to rotate the view. The applet and a more challenging one '*Building houses with side view*' are also available from the Freudenthal Institutes' Wisweb site together with a number of other useful applets (follow the 'applet' link).

## Objectives: Visualize and Work with Dynamic Images; Make and Justify Conjectures and Generalizations

### Manipulating in 2 Dimensions

One advantage of using dynamic geometry software such as *Cabri Geometry 2 plus, the Geometer's sketchpad* or *Geogebra* is that it allows you to drag points on geometric objects and to quickly be able to make and test conjectures and generalizations about properties which remain constant, such as angles which remain equal or shapes that retain their properties for example, parallel sides. One example of this is in exploring properties of angles in circles, and this can be seen in action in the classroom in the Circle Theorems lesson within the Teachers' TV programme 'Hard to teach secondary Maths using ICT' with additional information available from the 'Inspire me' section of the BECTa website.

**Further Resources**

There are many resources available to support teaching and learning using dynamic geometry software.

- A selection of files is available in the BECTa section of the CD 'Integrating ICT into the Mathematics Classroom' published by ATM
- The websites for dynamic geometry software *(Cabri Geometry II plus, The Geometer's Sketchpad, Geogebra)* all have additional help and support information
- There are two Teachers TV programmes which show dynamic geometry in action
    - KS3/4 Maths – Using Dynamic Geometry has classroom sessions showing how dynamic geometry software can help pupils better engage with learning mathematical concepts by interactive demonstrations, constructions and explorations and
    - KS3/4 Maths – Demonstrating Dynamic Geometry in which a Maths specialist teacher demonstrates a variety of ways in which dynamic geometry software can transform the teaching of mathematical concepts.
- Adrian Oldknow's website has many examples of dynamic geometry in action.

## Objective: Apply Suitable Mathematics Accurately within the Classroom and beyond; Identify the Mathematical Aspects of a Situation or Problem; become Familiar with a Range of Resources, Including ICT

**The 'Playground Maths' Project**

This uses images of roundabouts, swings and slides as contexts for work on a variety of mathematical topics. The initial trial of this activity with Year 10 was quite structured with specific tasks for each piece of equipment. The teacher went on later with a different class to set a more open task to *'Investigate the mathematics that you can find in a playground'*. A visit to the local playground gave students the opportunity to take digital photographs using cameras or mobile phones and to take measurements. The photographs were imported into Dynamic geometry software which was then used to measure distances, angles and gradients. Graphic calculators were also used for equations of linear graphs.

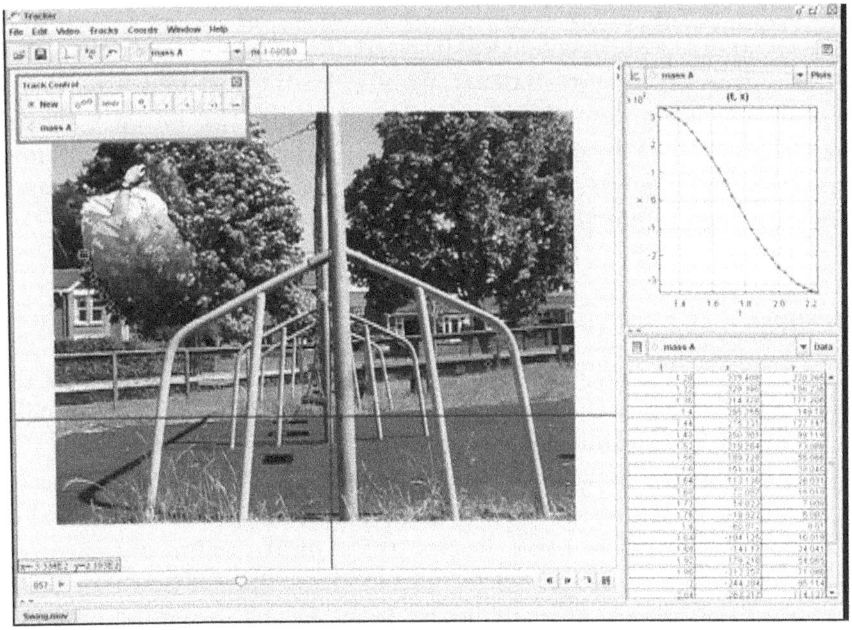

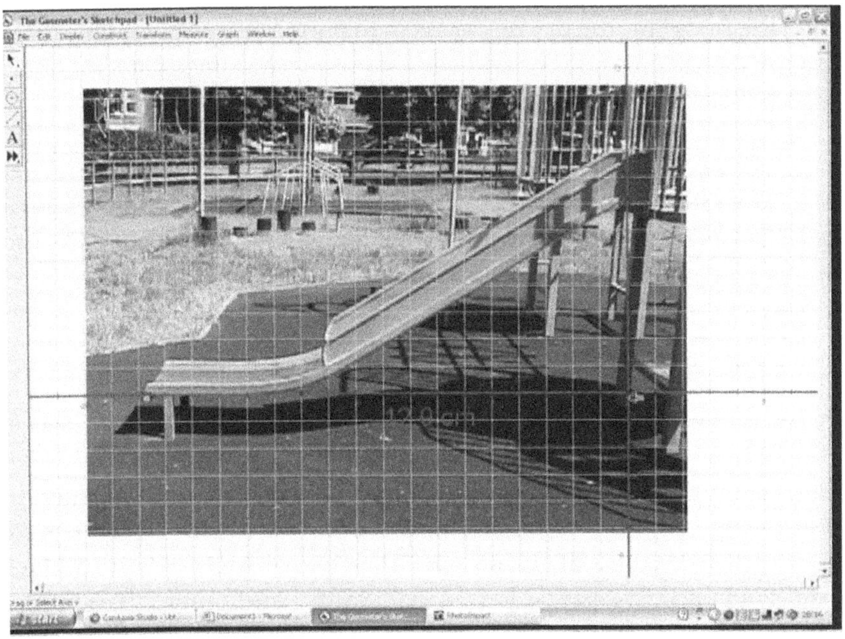

### *The Swing*

A video clip of a swing in action was inserted into open source video analysis software (Tracker 2) and students investigated the loci that a particular point on the swing traced out as the swing moved. The software allowed them to collect data for the coordinates of points on this path. A still, showing the coordinates, was pasted into dynamic geometry which was then used to find the locus of the path of the swing and the distance travelled.

### *The Slide*

An image of a slide was imported into dynamic geometry software and scales adjusted to match known measurements. This enabled other lengths, angles and gradients to be estimated.

The mathematics department in this large comprehensive school makes extensive use of the school's VLE to share activities between staff and to show students' work. The Year 10 class (set 3 of 5) produced a movie on DVD with the aid of the school technician. They showed how they could extract various aspects of mathematics from the playground equipment. They scripted the movie themselves and helped to edit it. The students also used screen capture software to make a video showing other students and staff how to use the video analysis software.

With the first class, the teacher was able to book a school computer room, while the video clip of the second class shows them using a class set of laptops which are available on a portable trolley.

This case study is one of those described in the QCA 'Engaging mathematics for all learners' project and there is also a short video clip about this activity 'Swings and Roundabouts' on the QCA curriculum website at First select 'case studies' and then 'mathematics'. A Teachers TV programme will also be available shortly.

## Some Additional Sources of Information and Resources

- Some of the initial images of slides were from the Problem pictures CD from Richard Phillip's website.
- There is also a selection of these images on the CD that accompanies the book *Integrating ICT into the Mathematics Classroom* published by the ATM (2005). The 'Maths Gallery' on this CD includes images of 'Straight lines and gradients', 'U-shaped curves' and 'Waves and other curves' which

can be inserted into dynamic geometry or graphing software so that students can experiment fitting graphs of functions.

- The MA website has a link to a booklet offering advice on maximizing the potential of the interactive whiteboard which includes using images of slides with IWB mathematical features.
- The free *Mathematical Toolkit* software also allows for modelling with both still and moving images and comes with a small library of both still images and video clips.
- Jing is free screen capture software which allows users to make short clips and to share them with others. More professional versions are available from links on the web page.

## Objective: Engage in Mathematics as an Interesting and Worthwhile Activity; Take Account of Feedback and Learn from Mistakes

### Distance-time Graphs

This is one of the lessons featured in the Teachers TV programme KS3/4 Maths: New Maths Technology – In the Classroom. In this programme a mathematics department embarks on a project to make better use of ICT in the classroom. The teachers are introduced to some of the new technology and then try out their newly-acquired skills. In particular, a lively Year 8 group has fun trying to recreate particular shaped distance time graphs time by moving in front of a data plotting range-finder (the Texas Instruments CBR2). This activity enables pupils to make direct links between their movements and the shape of the graph and to avoid many misconceptions which pupils traditionally have about these graphs. The single range finder can be linked to a graphical calculator with an OHP link or to a PC with software such as a graphic calculator emulator (TI84 emulator) or *TI Nspire* software.

### Further Information and Resources

- 'Real time distance time graphs' is one of the activities in the Practical Support pack's mathematics section
- Texas Instruments have a wide range of information, resources, research findings and classroom support materials available from their websites.

- Further information about this and other data logging activities can be found in the Becta book: *Data-capture and modelling in mathematics and science* which is available to download from page 2 of Adrian Oldknow's website
- The other lessons included in the TTV programme are:
  - Year 7 investigate algebraic expressions using an interactive number line. This is a free resource from Skoool available to download from the London grid for learning.
  - Year 9 explore rotational symmetry with dynamic geometry software
  - Year 10 investigate a problem using graphical calculators

## Objective: Work on Problems that Arise in Contexts beyond the School

Work with the handling data cycle: Specifying the problem and planning; collecting data; processing and presenting the data; interpreting and discussing the results

There is a wealth of data available on the internet which enables the making and testing of statistical hypotheses relating to large data sets without the time consuming process of collecting and organizing the data. The use of real world data and topical contexts adds a richness and relevance to the activity. One of the schools in the QCA 'Engaging mathematics project' made use of data from a fast food restaurant chain to make and test hypotheses relating to the fat, salt, sugar and calorie content of different categories of meals.

### Further Resources – Activities and Data Sources

- *QCA – RSS Centre Review of Handling Data and Statistics in GCSE Mathematics* consists of 8 case studies with teachers' notes and links to appropriate data sets
- The *Practical support pack* has a range of activities with teacher notes, lesson plans and resources. These include *'Population and development database'* and *'Wrist and neck sizes'*
- CensusAtSchool: is an International Children's Census collecting and disseminating real data for use by teachers and pupils in data-handling, ICT and across the curriculum for learning and teaching.

- Stats4schools: this website aims to help teachers and pupils to get more from statistics. For pupils, there are datasets to download and include in projects. For teachers, there are lesson plans and worksheets.
- Experiments at school: This website has a number of experiments to carry out and collect data which can then be analyzed. It includes a reaction timer. Further ideas and another reaction timer are on the NRich website.
- Further ideas for data-handling activities and sources of data can be found in the *'ICT enriched curriculum grid'* described earlier and available in the documents folder of the 'ICT and mathematics community' on the NCETM website.

## Objective: Recognizing the Rich Historical and Cultural Roots of Mathematics

This activity was developed initially by teachers with the help of internet research. Subsequently their students were given opportunities to do similar research on aspects that particularly appealed to them.

### The Golden Ratio Project

One school tackled an extended project on the Golden Ratio and Fibonacci sequence for the entire Year 10 cohort over a 2-week period. Details of this are included in the guidance booklet for the QCA 'Engaging mathematics for all learners' project. Two teachers prepared a variety of activities and resources. The teachers used ideas that they had collected from various sources including the NRich project website and Dr Ron Knott's website. There were more activities than the time available to give teachers and students the opportunity to choose between activities. They included: (1) a variety of constructions with instructions for doing them on paper or using dynamic geometry software; (2) statistical tasks looking for the existence of the Golden Ratio in the human form and (3) opportunities for students to do their own further research.

The students found the topic particularly intriguing and welcomed the opportunity of the further internet research to pursue topics that particularly interested them. They made good use of the schools VLE to post their ideas and comments.

One final note, the problem with citing links to websites is that they frequently change! The links were correct at the time of writing, but if any cease to work, then the 'Wayback Machine' often proves helpful in locating websites and resources that have disappeared.

## Websites

All websites accessed on 5 February 2011.

Adrian Oldknow's website http://www.adrianoldknow.org.uk/
ATM www.atm.org.uk/buyonline/products/rea025.html
Becta (Inspire me) http://schools.becta.org.uk/index.php?section=cu&catcode=ss_cu_ac_mat_03
Cabri Geometry II plus www.cabri.com/cabri-2-plus.html
Cabri 3d (software) www.cabri.com/download-cabri-3d.html
Cabri 3d (software overview video) www.cabri.com/bett-awards.html
Cabri 3d (diagonal of a cuboid lesson) www.cabri.com/cabri-3d.html
CensusAtSchool www.censusatschool.ntu.ac.uk/default.asp
Dr Ron Knott's website www.mcs.surrey.ac.uk/Personal/R.Knott/Fibonacci/
Experiments at school www.experimentsatschool.org.uk/main/
Freudenthal Institutes' Wisweb www.fi.uu.nl/wisweb/en/
Geogebra www.geogebra.org/cms/
The Geometer's Sketchpad www.dynamicgeometry.com/index.php
Jing www.jingproject.com/
Mathematical Association (Interactive Whiteboard advice) www.m-a.org.uk/jsp/index.jsp?lnk=140
Mathematical Toolkit http://lgfl.skoool.co.uk/common.aspx?id=901
NCETM www.ncetm.org.uk/
NRich (Golen ratio) http://nrich.maths.org/public/search.php?search=Golden+ratio
NRich (Reaction timer) http://nrich.maths.org/public/search.php?search=Reaction+timer
Practical Support Pack www.teachernet.gov.uk/wholeschool/ictis/cpd/practicalsupportpack/
Problem pictures (Richard Phillips) http://www.problempictures.co.uk/index.htm
QCA – RSS Centre Review of Handling Data and Statistics in GCSE Mathematics www.rsscse.org.uk/qca/resources0.htm
QCA (curriculum website) http://curriculum.qca.org.uk/key-stages-3-and-4/index.aspx
Skoool http://lgfl.skoool.co.uk/common.aspx?id=901
Stats4schools www.stats4schools.gov.uk/default.asp
Teachers' TV (diagonal of a cuboid/Circle theorems lessons) www.teachers.tv/video/29853
Teachers' TV (KS3/4 Maths: Demonstrating Dynamic Geometry) www.teachers.tv/video/3080
Teachers' TV (KS3/4 Maths: New Maths Technology) www.teachers.tv/video/154
Teachers' TV (KS3/4 Maths: Using Dynamic Geometry) www.teachers.tv/video/3081

Texas Instruments http://education.ti.com/educationportal/sites/UK/homePage/index.html
Tracker 2 www.cabrillo.edu/~dbrown/tracker/
Wayback Machine www.archive.org/index.php
Yenka http://www.yenka.com/

## References

ATM. (2005). *'Integrating ICT into the Mathematics Classroom'* ( Derby, UK, Association of Teachers of Mathematics www.atm.org.uk/buyonline/products/rea025.html (accessed 5 February 2011).

DfES. (2005). *Improving Learning in Mathematics,* Standards Unit http://tlp.excellencegateway.org.uk/default.aspx#math_learning (accessed 5 February 2011).

QCA. (2009). *Engaging Mathematics for All Learners* www.qcda.gov.uk/22223.aspx (accessed 5 February 2011).

Chapter 15

# The Uses of Online Resources for Teaching and Learning Mathematics at Advanced Level

Bryan Dye
*Consultant*

## Introduction

In this chapter we look first at the background to online learning, secondly at the kinds of tools and technologies available online, thirdly at how three particular websites are providing resources for A Level mathematics, and lastly at the intended users of these resources.

## Background

Teenagers nowadays are sophisticated users of the internet. In the United States four years ago it was estimated that about 87 per cent of teenagers between 12 and 17 were online. The *Guardian* newspaper recently reported on research revealing that 65 per cent of teenagers are streaming music regularly, with more 14- to 18-year olds (31%) listening to streamed music on their computer every day compared with music fans overall (18%). The BBC is promoting its online Iplayer concept. In January this year 26 per cent of 14- to 18-year olds admitted filesharing at least once a month. Older people among us may still be wondering what 'filesharing' actually is. The point is that teenagers will expect a lot from their internet resources and take a lot for granted. A few helpful notes, even an online textbook is no good; interactivity is the key, ranging from video clip and games to fully controllable functionality. In any case, it can be argued that the textbook, having once been scanned into a computer and then left for dead, has now evolved into the web phenomenon that is the wiki. According to Wikipedia, a **wiki** is a type of collaborative software program that typically allows web pages to be created and collaboratively edited using a common web browser. Websites running such programs are themselves referred to as wikis. A search on the web for information on virtually any topic will usually throw up a link to a wiki within the first half dozen references. This is what young people are becoming accustomed to; interactivity to them no longer has any 'wow' factor, it is expected.

## Online Interactivities

The variety and degree of interactivity available on-line is increasing steadily. Whereas a few years ago educational sites may have offered little more than text, now there is a myriad of interactivity utilizing technologies such as Flash, Java, Javascript, PHP, forums, conferencing and video. Recently the browser Firefox, in version 3.5, has enabled the embedding of video directly into a web page with out the need for any special display software. The web is rapidly approaching the point where full interactivity is accessible to all. So, what is available to the advanced level mathematics student? Mathematics seems to me to be ideally suited to the web. The student can investigate graphs, geometrical constructions, algebraic manipulation, logical thinking, calculus, mechanics, statistical diagrams, simulations, video, and so on. Some of these interactivities have been born and bred solely on the web, interactive geometry being a good example. On the one hand proprietary software such as **Cabri Geometre**, **Geometer's Sketchpad** or **Autograph** have developed their own on-line versions; on the other hand new packages

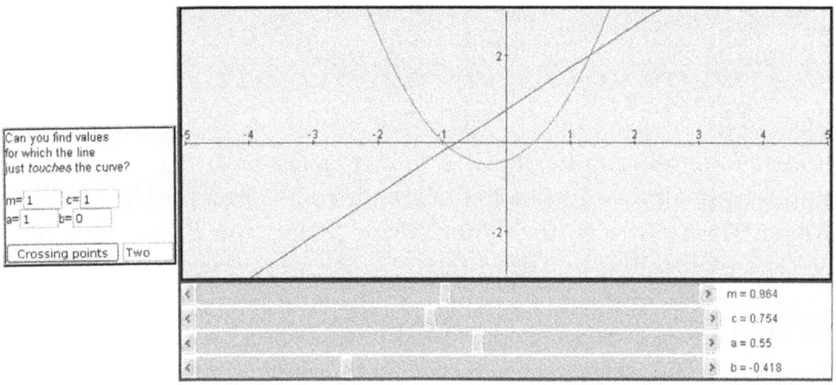

such as **GeoGebra** (have come about purely on the web. In the relatively short time that it has been around, Geogebra has integrated algebra with geometry and recently spreadsheets, developed its own support forum and its own 'wiki'. Organizations like The Association of Teachers of Mathematics and the Further Maths network in this country and the Mathematical Association of America in the USA are looking closely at what this kind of software can offer. Other online interactive packages include **JavaMath** or a recent update of it **WebCompMath** (both of which present the user with interactive graphs whereby the effect of a change in parameters (using an slider) on the shape of a curve may be observed. The graphs may be polynomial, parametric or polar. Differentiation, integration or composition of functions can be investigated also in this same environment. Online spreadsheets have come of age with **Zoho** (http://sheet.zoho.com/)

The correct display of mathematical notation has in the past been an issue for web authors, whereby fractions, indices and matrices have been displayed in a primitive, even amateurish form such as **3^2**, or **4/5**. Now there are web typesetting technologies such as **TeX** which can be incorporated into web pages to enable precise display of mathematics, for example:
Definite integration:

$$\begin{aligned}\int_1^5 3x^4 + 2x\,dx &= \left[\frac{3}{5}x^5 + \frac{2}{2}x^2\right]_1^5 \\ &= \left(\frac{3}{5}5^5 + \frac{2}{2}5^2\right) - \left(\frac{3}{5}1^5 + \frac{2}{2}1^2\right) \\ &= 1900 - 1.6 = 1898.4\end{aligned}$$

Multiplication of matrices:

$$\begin{aligned}\begin{pmatrix}-4 & 0 \\ 3 & -2\end{pmatrix}\begin{pmatrix}3 \\ 4\end{pmatrix} &= \begin{pmatrix}(-4\times-3)+(0\times-4) \\ (3\times-3)+(-2\times-4)\end{pmatrix} \\ &= \begin{pmatrix}12+0 \\ -9+8\end{pmatrix}\end{aligned}$$

For the technically minded the TeX code **-b \pm \sqrt{b^2 – 4ac} \over 2a** would be displayed as the familiar quadratic roots formula.

**Adobe Flash** offers further interactivity. In fact the full integration of video, animation, drag and drop and other mouse actions is possible. Some websites have maximized this potential, MyMaths being a prime example. MyMaths uses a sophisticated Flash-driven display to take the student through a topic in stages. At each stage the student is encouraged either to follow a train of thought on-screen or else to interactive with it to solve

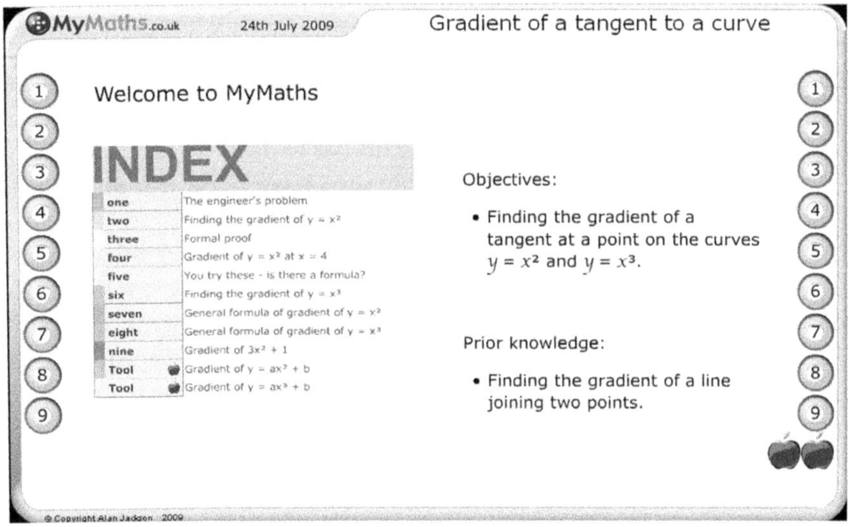


problems or submit numerical solutions. It might be argued that to some extent the MyMaths content is designed to replace the teacher.

In the United States the site **Explorelearning** has for some time been using Flash and Shockwave to develop interactive material for pupils in grades 3 to 12, not only for mathematics but science too. Mathematics is covered up to the level of college algebra and pre-calculus. Explorelearning call these interactive resources 'gizmos' and although the site is not free, many of the 450 gizmos available can be viewed without payment for a few minutes.

## A-Level Mathematics Websites

So, there is a lot out there, and it gets better all the time. But one problem remains: access. How does the busy teacher or student find this stuff when they need it? Many sites on the internet have in the past specialized in providing links to recommended sites, but such lists, which are often extremely long, are hard to maintain and very hard to ensure continued relevance. As far as A level mathematics goes, there are three sites in the United Kingdom that are attempting to provide a coherent approach to managing a large body of interactive material. They are:

1. MathsNetAlevel
2. Further Mathematics Support Programme (FMSP)

3. MyMaths

None of these sites provide their content completely for free, but require yearly subscription. MyMaths is already well known in schools for its coverage of GCSE mathematics and has within the last year begun developing A-Level modules. The other two have been established for some time.

A difficulty confronting online providers is the various examination specifications available in the United Kingdom. Currently there are those provided by AQA, Edexcel, OCR, OCR-MEI, CCEA and WJEC. Scottish Highers and Advanced Highers are significantly different in style and content to the rest. Though the specifications are similar they are not identical. OCR-MEI, for example, offers modules that the others do not; 'Differential Equations' and 'Numerical Methods' being two of them. Both MathsNetAlevel and the FMSP cater in detail for these various specifications, while MyMaths is part of the way there.

FMSP and MyMaths offer a subscription within which the user chooses which syllabus elements they want to use. MathsNetAlevel requires the user to choose their syllabus at the outset. Once the user has logged on, they must select the module they wish to work on and then choose a topic. From there on a vast array of resources awaits. Here is a selection from the three sites:

Step by step illustration of algebraic techniques:

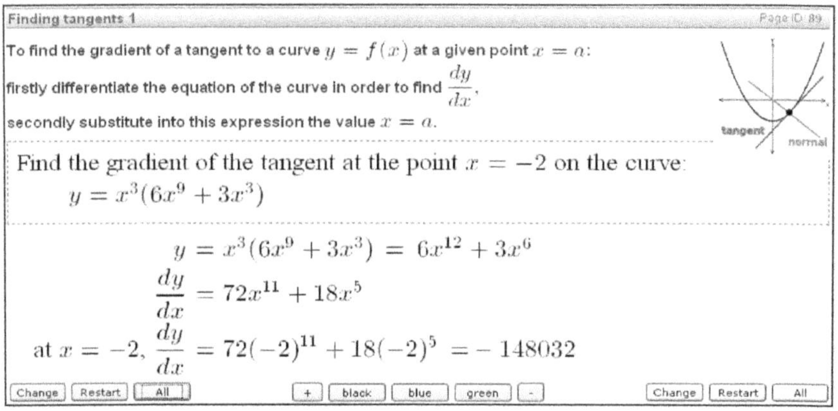


Investigation of quadratic graphs

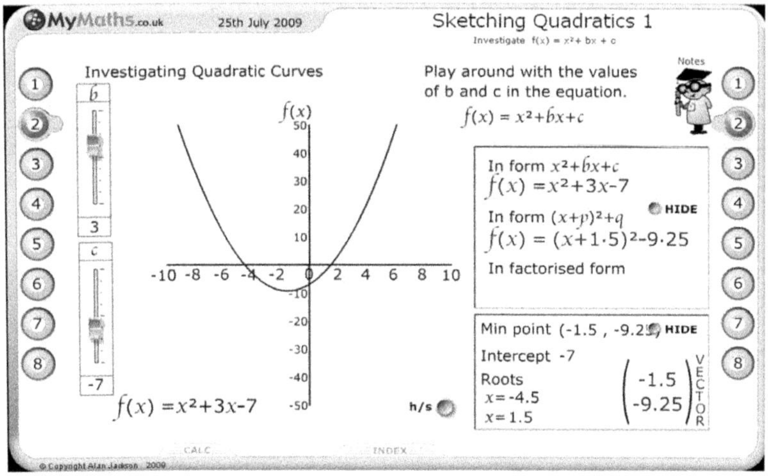


Investigate parabolas

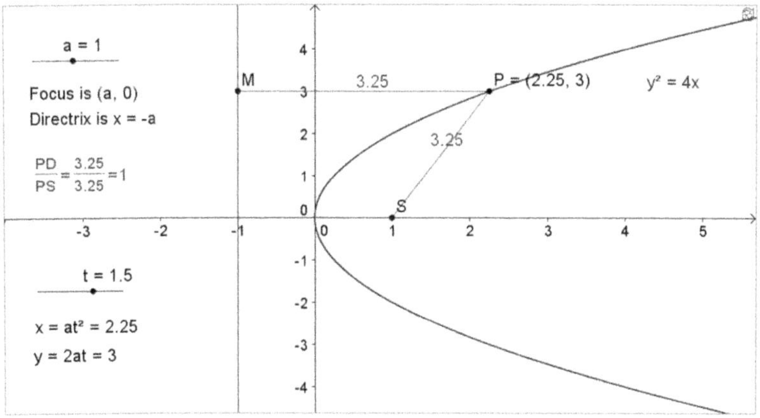


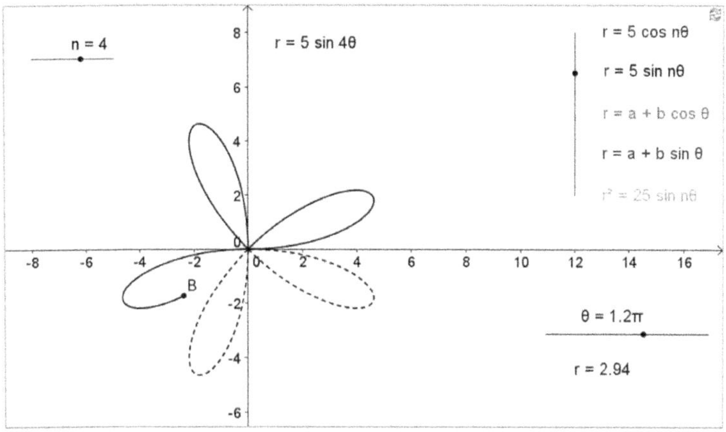


Investigate polar graphs

## The Users

The above websites all aim to support the student in their studies at home or at school and are also intended to be used by teachers in the classroom with the aid of an interactive whiteboard. The resources are multipurpose. Here are four possibilities.

### The Student

The student sitting alone at home can work through carefully structured sequences of work, where feedback is usually immediate and on-screen. The student can, to a large extent, teach themselves. This may become increasing important given the problem that schools are finding in funding further mathematics courses in their establishments – and in some cases AS and A2 courses too. There is consequently a distinct rise in the number of students following advanced level courses by 'self-study'. One of the fundamental jobs of the Further Mathematics Support Programme (FMSP), which came into existence in August 2009, is to support schools and colleges in a similar way to the previous Further Maths Network. The aim is to help schools and colleges that do not currently offer Further Mathematics to start offering it; provide CPD courses to help schools and colleges to develop their capacity to teach Further Mathematics; and liaise with NCETM to establish networks to support teachers of Further Mathematics. 'Self-study' is unfortunately still the only route some students find to a further mathematics A-level in regions where the FMSP has not got off the ground.

### The Teacher

These interactive tools are also aimed at the teacher. They are usually designed to work with minimal technical knowledge required on the teacher's part, and with an interactive whiteboard in mind. Mouse actions such as drag and drop are heavily utilized. The tools are not in general open-ended but designed to illustrate or solve a specific mathematical problem, with the advantage that the teacher does not spend time learning how to run the software but can instead concentrate on how to integrate it into his or her lesson planning. The teacher can build their lesson around these tools, knowing that the student will then be able to follow up the lesson themselves in their own time using the same resources. One of the new features of MathsNetAlevel is the ability to create one's own randomized exam paper. Questions can be freely chosen from a specific module and then printed out along with a corresponding mark scheme. The teacher can issue the paper as a homework task, or else require the students to print their own, to be followed up by a formal test in class. The randomizing of the question content means that repetitive practice on the skills needed is fully realized, and relevance is maintained by regular updates to the site based on the most recent public examinations.

### The Tutor

Besides the student and their teacher, a third potential aim of these activities is the online tutor serving the function of the conventional teacher only without the face-to-face element. The internet provides forum and conferencing facilities, even free online whiteboards such as 'scriblink' or 'skrb', all of which could be combined with on-line phone facilities such as 'Skype' to create a fully communicative medium. With these technologies, the online tutor can talk and write to their students, illustrate the topic under discussion and point their students towards follow-up work, and they can use a forum to keep this all organized with dates for meetings and deadlines, and so on.

## Conclusion

### Professional Development

The ideas summarized for the tutor can be utilized further by groups of teachers for the purpose of professional development. As mentioned earlier a issue confronting the provision of A Level courses is the number of students applying but also there is the issue of the ability and suitability of teachers to teach the course, particularly as specifications change. Using the tools of forums, conferencing, collaborative online whiteboards, the likes of Skype and the resources provided by the A-level mathematics websites, groups of teachers can receive professional training in their own homes.

Teachers can thus develop and improve their mathematics understanding and at the same time become familiar with the kind of online interactions that their students will already be taking for granted!

## References and Websites

All websites accessed on 5 February 2011.

Autograph www.autograph-math.com/
Explorelearning www.explorelearning.com/
Further Mathematics Support Programme: www.furthermaths.org.uk/index.php
Geogebra www.geogebra.org
Geometer's Sketchpad www.dynamicgeometry.com/
JavaMath http://math.hws.edu/javamath/
MathsNetAlevel www.mathsnetalevel.com
MyMaths www.mymaths.co.uk
NCETM www.ncetm.org.uk/
Scriblink www.scriblink.com/
Skrbl www.skrbl.com/
Skype www.skype.com/intl/en-gb/
Weompmath http://webcompmath.sourceforge.net/

## Chapter 16

# What Do the Subject Associations Offer?

Ruth Tanner
*Mathematics Consultant, Shropshire, UK*

## Introduction

Most teachers of mathematics spend some time learning about ways to use Information and Communications Technology (ICT) to enhance their teaching during their training. However, their potential to explore and practise the ideas that they encounter or develop depends, to some extent, upon the facilities available at their placement schools. Once qualified and working in a school it becomes much harder for teachers to continue developing and extending their use of ICT, unless they are fortunate enough to work alongside colleagues who are skilled users of ICT.

Understandably therefore, many teachers of mathematics use the expensive ICT equipment available to them as an electronic text book or to support their classroom management (e.g., by displaying objectives, notes and homework tasks using pre-prepared PowerPoint slides or interactive whiteboard flipchart pages). In contrast to this, the document, 'Secondary mathematics with ICT – A pupil's entitlement at Key Stages 3 and 4'[1] outlines 'six major opportunities for learners to benefit from the use of ICT in mathematics'. This inspiring document was produced by Becta in association with the Mathematical Association (MA)[2] and the Association of Teachers of Mathematics (ATM)[3]. Learners whose teachers consistently limit their use of ICT to little more than displaying notes, however colourful, will miss out on the engaging and stimulating teaching approaches that this document suggests ought to be an entitlement for learners of mathematics in the twenty first century. Developing and improving their use of ICT can be difficult for some teachers. In my own journey to become an ever more effective and innovative user of ICT the mathematics subject associations, and in particular the MA and ATM, have always played a vital role and they still continue to do so.

## Subject Association Journals, Publications and Software

Teachers who are members of the MA and ATM regularly receive the subject association journals (*Mathematics in Schools, Primary Mathematics* and *Mathematics Teaching*) and since July 2009 members of ATM can also access the on-line, interactive journal Mti[4] (*Mathematics Teaching interactive*). These journals offer teachers inspiration, challenge, support and encouragement as they seek to reflect on their teaching and explore new ideas, including those that involve the use of ICT. Software packages such as graph plotters, dynamic geometry and computer algebra systems have so much potential to enhance the teaching of mathematics, but understandably teachers often find it hard to know how to introduce these packages into their classrooms. Reading accounts of the use of these packages in *Mathematics Teaching, Primary Mathematics* and *Mathematics in Schools* can be a good way to acquire ideas and gain the confidence needed to get started. Many articles are now supported by online resources, especially those that appear in MTi.

An article may teach a new technical skill. For example some teachers will have been introduced to constructing sliders in Excel for the first time when reading the *Micromath* article *Scroll Bars in Excel.*[5] Articles have the potential to open up a whole new area of use for a software package that the reader is already familiar with. For example, I still remember how reading Adrian Oldknow's article 'Mathematics from still and video images'[6] introduced me to the idea of pasting photographs into dynamic geometry or graph plotting software to encourage work on transformations. This then led me to experiment with pasting local maps instead of photographs into dynamic geometry software. Thus creating a tool for exploring bearings in a more meaningful context than ships and lighthouses for learners in the Midlands! But above all, along side the lesson ideas and technical tips, these articles nearly always focus on the pedagogy of using ICT and the ways in which its use can open up new approaches to learning some aspects of mathematics. The focus moves from the teacher getting the equipment connected and then running the software correctly to the pedagogy involved in improving learning.

As it becomes more and more difficult for schools to release teachers to attend courses during the school day the journals of the subject associations have the potential to be used to offer an alternative form of professional development. Teachers working individually or in pairs could choose an article which is relevant to them and experiment with some of the ideas in their own classrooms. Indeed, working in this way could become part of an individual teacher's performance management or a whole department development plan with appropriate time and resources allocated to it.

In addition to the journals the subject associations sell books and resources. ATM also sells some software of its own which it claims is designed 'to encourage discussion, support teaching and challenge students' thinking'.[7] Because of these design aims the products are accompanied by support materials and lesson ideas to guide less experienced users of ICT. Teachers who use some of these support materials will usually find that, as well as their students being challenged to think, discuss and reason they themselves are also challenged to think about their pedagogy. Consequently the teacher becomes more effective at exploiting the potential of ICT in the classroom.

## Subject Association Branches and Conferences

For many teachers their contact with any subject association that they belong to is solely through the journals and publications. These teachers are missing out on another important aspect of the work of these bodies which is the provision of both national and local conferences and the work of local branches.

Over recent years many subject association branches have included meetings and workshops on aspects of teaching and learning mathematics using ICT. Branch meetings generally take place two or three times a year, out of school hours (often on a Saturday morning or a weekday evening) but the extra commitment required to attend is always well worth it! The ethos of a branch meeting is very different to that of a commercial course led by a paid tutor. A branch meeting or a conference brings together teachers with a variety of experiences from a wide range of educational establishments and phases (I have seen a newly qualified teacher working on some mathematics with a retired Ofsted inspector at a branch meeting.). Being able to see ideas in action and talk to those enthusing about them is even better than reading about them. It is always good to see equipment or software being used by a colleague and have the chance to try it out oneself. The result is usually an added enthusiasm and determination to pursue the funding required to purchase the new equipment or software. I remember talking to a teacher who had a data projector but had been waiting for

2 years for an interactive whiteboard for her classroom. At a branch meeting she saw a tablet PC, wirelessly connected to a projector, being used as an alternative. When she got back to her school the network manager told her that the projector in her room already had the wireless facility and that she could swap her laptop for a tablet PC. Less than a month after the branch meeting she had the new equipment up and running and was thrilled.

It is often when software or hardware is used at a workshop that teachers become inspired to have a go themselves. For me this was my way into getting students to use dynamic geometry. I had taught myself to use some dynamic geometry software I had purchased but it was only after attending an ATM workshop where we used dynamic geometry to explore the properties of quadrilaterals that I began to use dynamic geometry with students.

## Subject Associations Working with Other Agencies

The Mathematical Association and the Association of Teachers of Mathematics frequently work with each other and with other agencies such as British Educational Communications and Technology Agency (Becta), the Department for Children, Schools and Families (DCSF)· the National Centre for Excellence in Mathematics Teaching (NCETM)[8], and Bowland Maths[9] to produce resources and provide professional development aimed at empowering teachers to make better and more effective use of ICT. These partnerships have resulted in some excellent free resources for teachers in the United Kingdom.

## Micromath Book and CD

One such collaboration resulted in a free book and CD entitled *Integrating ICT into the Mathematics Classroom*[10] being given to every maintained secondary school in the United Kingdom. In 2005 *Micromath,* an ATM journal, merged with *Mathematics Teaching* and the book is a celebration of the 21 years of the journal's contribution to encouraging and developing the use of ICT in classrooms. The CD is well worth looking at if you can find it. Not only does it contain a more extensive archive of articles than the book but it also contains a wealth of other free resources. The CD contains a gallery of 125 photographs for use with an interactive whiteboard, dynamic geometry package or graph plotter; a collection of free software; some excellent tutorials for getting started with Cabri and Geometers Sketchpad and some resources for graphical calculators. Also on the CD are six units of work using ICT. Although these units were developed prior to the new National Curriculum they could be used to support the teaching of the Key Processes set out in the New Curriculum[11].

## Practical Support Pack

The DCSF website has a section devoted to learning and teaching using ICT called the Practical Support Pack[12]. The secondary mathematics section of the Practical Support Pack offers another collection of units of work which involve the use of ICT that were developed with the support of the MA.

## Teachers' TV

The subject associations continue to work closely with Teachers' TV[13] which provides a range of free videos giving practical examples of ICT being used in mathematics classrooms and of individual teachers and departments increasing and improving their use of ICT.

## Delivering Professional Development

From time to time the subject associations also work with some of these other agencies to provide 'hands on' professional development to help to build up the confidence of less experienced users of ICT, introducing them to some practical ideas and giving them the time to try things out in a supportive environment. The evaluations from these events invariably show that most teachers return to their classrooms feeling more confident and enthusiastic about trying out some of the ideas they have been working on. Again these events give delegates a chance to try out not just software but also equipment that may be new to them and an opportunity to see the use of ICT as more than just interactive whiteboards.

## Conclusion

It is not necessary to know everything about a piece of software in order to be able to begin to use it effectively and creatively. Most open software is so powerful that few of us know its full potential, there is always more to learn!

The software and hardware that teachers have access to are constantly developing and improving.

One of ATM's four Guiding Principles is:

> It is important to examine critically approaches to teaching and to explore new possibilities, whether deriving from research, from technological developments or from the imaginative and insightful ideas of others.

Through the journals, branches and conferences some members of the subject associations continue to 'explore new possibilities' and to learn from the 'insightful ideas of others 'as they strive to use ICT to improve the experiences of twenty first century learners of mathematics.

## Notes

All websites accessed on 5 February 2011.

1 www.teachernet.gov.uk/teachingandlearning/subjects/ict/bectadocs/sec/
2 www.m-a.org.uk
3 www.atm.org.uk
4 www.atm.org.uk/mti
5 Mulkerrin, P. (2000) 'Scroll bars in Excel'. *Micromath*, 16/2, 24–27.
6 Oldknow, A. (2000) 'What's in it for Mathematics?' *Micromath*, 16/2, 30–34.
7 Association of Teachers of Mathematics, *Mathematics Resources Catalogue 2009–2010*, pp. 8–9 www.secondarymathsite.co.uk/Organisations/ATM/ATM1/ATM_Web_Catalogue%20Final.pdf
8 National Centre for Excellence in Mathematics Teaching. www.ncetm.org.uk
9 www.bowlandmaths.org.uk
10 Edwards, J. and Wright, D., *Integrating ICT into the Mathematics Classr*oom, Derby, Association of Teachers of Mathematics 2005
11 http://curriculum.qcda.gov.uk/
12 www.dcsf.gov.uk/psp/index.aspx
13 www.teachers.tv

Chapter 17

# Modelling, Functions and Estimation: A Pizza Problem

Chris Olley
*King's College, London, UK*

## Introduction

As part of the Bowland Trust project to produce teaching materials to support applications of mathematics, King's College, London, developed a project in which students watched pizzas cool. The temperature of the pizzas is measured over time using data logging apparatus.

The pizza shop owner has hired mathematical consultants to support them in maximizing their market and hence their profit. Pizzas need to be developed fresh and hot. Many issues need to be addressed in this project, but two key questions emerge: How far can my deliveries reach with the pizzas remaining sufficiently hot? Does it matter how I package them? This requires two experiments, first to determine the least acceptable temperature, and secondly, to determine the time taken to reach this temperature with different packaging. The first is simple and requires a volunteer prepared to eat small pieces of pizza as it cools. Generally, the volunteer is not hard to find! It is the second experiment which throws up a lot of interesting mathematics.

It is important to recognize that there is no pizza shop, nor will our report actually affect one. Our students are principally motivated by a desire to get on in their mathematics. Frequently scenarios are presented in class as problems being solved as if they were in the 'real world', that is, motivated by the requirements of the supposed scenario. Paul Dowling (1998) refers to this as the myth of reference. Clearly, pizza shop owners are unlikely to employ consultants with data logging apparatus and computer algebra systems for advice on their business plan. However, the scenario brings a welcome sense of fun, plus the warm smell of pizza to the mathematics classroom. This recognition is important. Details of the mathematical development could be argued as unnecessary for scenario and the mathematical development would be restricted. We need to retain our perspective on our real purpose, being to develop the mathematics. We do nonetheless have a setting which motivates the desire for quality outside our area of expertise. The two experiments require some serious scientific consideration. Indeed, when we have presented this idea to adult teacher professionals, the focus has often been on critiquing the quality of the experimental set-up. Students in our trial schools did not. However, it clearly provides an opportunity to collaborate with science departments, who would wish to discuss the experimental design and improve it. Equally, an initial discussion about maximizing the profit of the pizza business throws up many issues (e.g., total profitability, cost/benefit of additional employees, food quality issues, etc.) which are outside the experience of most mathematics teachers. Instead of making an unrealistic attempt to 'deal' with them these could be studied in the business studies department, where this expertise resides.

What do students see when they look at a graph? Activities involving story graphs are frequently designed to help the student visualize the change in one variable dependent on another. However, constructing the story is difficult and the simplifications in the graphical representation often strain credulity – the strange stories of children walking with constant speed being a case in point. I have frequently used real time distance logging apparatus with experienced teachers and have been struck by how often they walk towards it when the distance/time graph they are supposed to be tracking goes up or ask me where they should start when the graph clearly shows the distance at time zero (see, Teachers TV, 2006). That there is a complicated link, often weak, between the scenario and the graph seems clear. Having, the possibility to collect the data in real time as a clearly known process is unfolding in front of the learners' eyes (and nose!) provides a powerful link.

Jeremy Rochele has developed SimCalc, a software simulator which produces cartoon images of motion activities (e.g., characters running at different speeds) together with a graph and table of values. This provides an example of multiple representations (see Ainsworth and VanLabeke, 2004) which is a key design precept of the TI-nspire software discussed in this paper. Rochele et al. presents outcomes from control and treatment groups involved in using the SimCalc to study 'the mathematics of change and variation'. They found

> Although, on average, Treatment and Control group students progressed equally well on simple mathematics, the Treatment group gained more

> on complex mathematics. For example, at post test, Treatment students were more likely to use the correct idea of 'parallel slope as same speed', whereas Control students were more likely to have the misconception 'intersection as same speed'. (Roschelle et al. 2007)

Test items examined exactly the misconceptions that SimCalc is designed to address, so it would be interesting to compare the quality of the input received by the control group. Nonetheless, this seems to have been effective in generating a felt link between motion and graph. In our case the measured change is happening in reality (rather than virtual reality) which may perhaps create a stronger link, although this remains to be examined.

It is necessary to be clear about our purpose here. We intend to watch the change in temperature of a pizza over time, in order to find out how long it takes to reach a certain value. (In our experiments we found 48° to be the least acceptable temperature). Now this could take a long time, longer than we could reasonably continue measuring for (certainly in an ordinary lesson). So, we will see if we can find a rule for the rate of cooling that will enable us to predict how long it will take. The predicting aspect requires the setting up and critiquing of a mathematical model, yet seems sensible enough in the context of the scenario. It is routine in classrooms to ask students to estimate. However, students need to have the opportunity to develop their skills in estimation and critically reflect on the how they estimate the future temperature.

In the classroom we tried two different models. One featured the teacher controlling the experiment using one microwave oven placed at the front of the classroom with one probe and computer set-up. The second featured 6 groups of students taking turns to use one of two microwave ovens at either end of the room, each group having its own probe and computer. The data logging equipment produces a real time graph which shows how the temperature is decreasing with time. It also shows the temperature on screen. We produced a worksheet in which students are asked to predict the initial temperature (actually the point at which the temperature starts decreasing, to take account of the probe heating up). Next to each prediction is a space to state the basis on which the prediction was made. Initially this will be due to external factors (guess, how hot ovens get, etc.). As soon as the pizza comes out of the oven it was placed inside one of the packaging types and the probe inserted. (Mini deep pan pizzas were used to ensure that the temperature of the topping was being measured, rather than the base). Students had already made their prediction for the peak temperature (i.e., time zero), so the data logger was set running. Immediately students are asked to

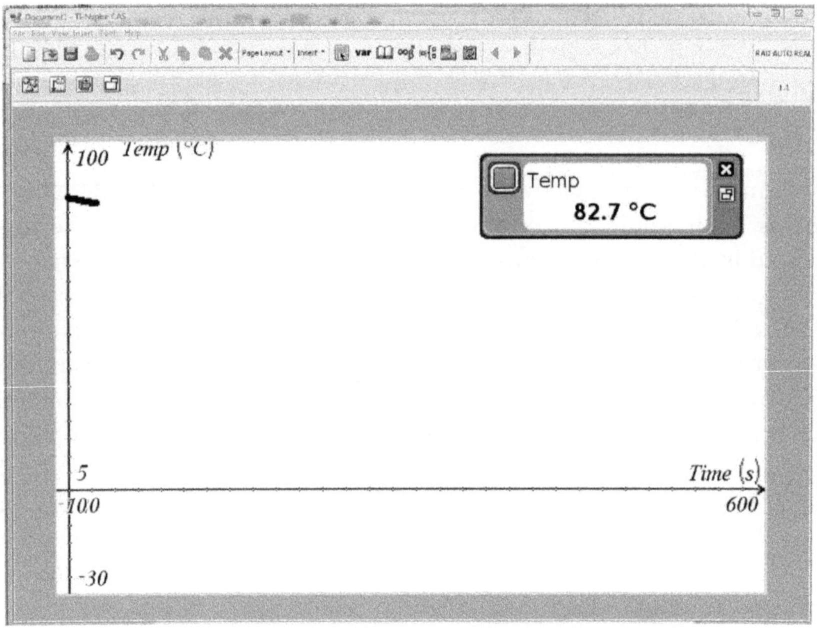


predict the temperature at the end of the 1st minute. The experiment is timed (we used a volunteer timekeeper to shout out 5-4-3-2-1 at the end of each minute). At the end of the 1st minute, the actual temperature is logged on the worksheet and a prediction is made for the end of the 2nd minute. Now, some students start to look at the rate of cooling as an indicator to support their estimate for the end of the 2nd minute. Also, students are asked to estimate the temperature at the end of the 5th minute. The experiment continues in this way for 10 minutes, each time students make future estimates. After 5 minutes they estimate for end of the 6th minute and the end of the 10th minute. By now, students are taking close account of the rate of change and using this to make better and better estimates for each successive minute. There is an aura of quiet competitiveness and satisfaction when estimates are close or even perfect. (Reading are taken to one decimal place). After 10 minutes, they estimate for 30 minutes, 120 minutes and 24 hours. Again the requirement to explain the basis for the estimation is emphasized. This last part requires students to share their mechanisms for estimation and discuss how they expect the temperature to change after the experiment has ended, that is, into the never-to-be-known. It was gratifying that students happily watched a pizza warming in a microwave oven for 2 minutes then watched it cooling for 10 with rapt attention!

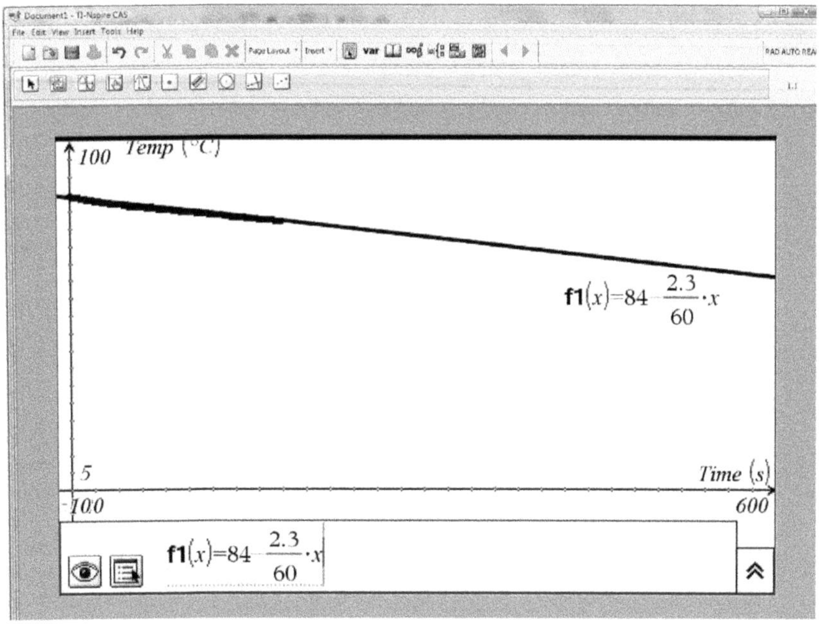


At the end of the experiment it is clear that the pizza is nowhere near down to the 48° minimum, so we need some way to work out when it will reach that temperature. The stage is now set for the key conversation: on what basis were the estimates made. Typically the cooling graph looks very linear. When asked how they estimated most considered answers were along the lines of 'it was going down about 2.3° a minute' or sometimes 'for the first five minutes, it was going down about 2.4° a minute and then for the next five minutes, about 2.1° a minute', Both clear statements of linearity. Depending on the available equipment, the students either drew a graph by hand from the data on the worksheet, or had access to a dynamic graph within the data collection software. By tradition, a line of best fit seems an obvious thing to make, so the possibility of setting up a model comes out naturally. Starting with the simpler suggestion of 'going down 0.8° a minute' we can ask, so what was the starting temperature? The software allows us to enter a function, to fit the data. So we start at the peak temperature (in the example it is 84°). So starting with $f1(x) = 84$ makes sense. Then it went down by 2.3° per minute so we make it $f1(x) = 84 - 2.3x$. The set up makes this look very natural. But then when we hit return to draw our best fit line something is clearly wrong. It requires very little prompting to see that the 2.3° was per minute, but the data was gathered

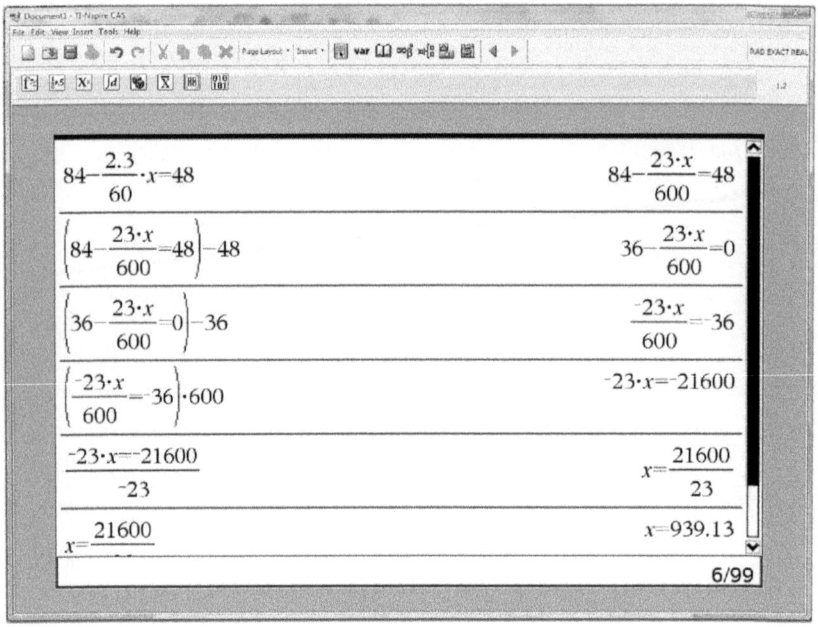


per second. So we can edit the function to show f1($x$) = 84 − (2.3/60)$x$. Happily the software gives an immediate response, so testing different theories for accommodating the minutes to second conversion can be done quickly by test and check. This feature keeps the discussion on track and avoids being sidelined by tricky numeracy issues. These can be returned to later. Often the very beginning of the experiment has an uneven cooling rate, so a little 'tweaking' of the model needs doing. Having seen the construction of the model, students feel in control of the coefficients. They can move it up or down a bit by changing the 'starts at' value and change the steepness by varying the 'per minute' value. They are very impressed by their capacity to make a near perfect fit. Looking at the graph of our best fit function, we can see roughly when the temperature will be down at 48°, though this may require extending the axes. However, the estimating power of the function becomes clear immediately. Students who hand drew their graphs immediately see the flexibility of the software. Fitting a graph by eye is very useful as it reinforces the power of the function. Clearly this function fits our data very well and so we can use it to find when the value of this function is 48. That is 84 – (2.3/60)$x$ = 48. Immediately students recognize that this is an equation. Their knowledge of how to solve it can now be brought to bear. Powerfully the software includes a computer algebra system (CAS). Here we can state the equation. Then work on it in

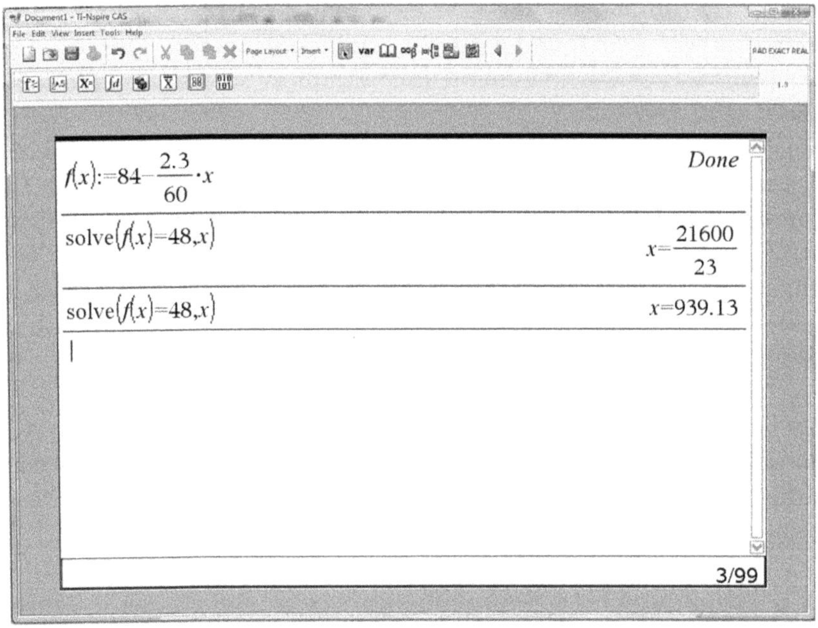


whichever way students suggest. Sensible and not so sensible suggestions can be tested and their outcomes considered. That there are many routes to solution is very empowering here. In the CAS we simply type the equation and enter it. Then take simply state what we wish to do to both sides. (CAS rightly cannot accept a fraction of 2.3/60 and so writes it correctly as 23/600, generating another key intervention). In the example, we took away 48, then took away 36, then multiplied by 600, then divided by −23 (a route suggested by a student). The CAS shows the result of an operation applied to the whole equation. This is quite a striking notion and has subtle advantages over the 'both sides' argument. That we have taken away 48 from the whole equation is more resonant with ideas of equivalence between statements. This does give us the potential for an interesting discussion later. This is about 15 and ½ minutes. For more complicated functions we may not (yet) have the tools to find a solution, so it is useful to demonstrate the solve function in CAS. We simply define a function $f(x){:=}\,84-(2.3/60)x$. (Note that we use the symbol ":=" to mean "is defined as". The different uses of the equals sign are frequently glossed over in classrooms and remain a key source of algebraic confusion for students.). Here we are forced to recognize the difference between the function definition and the equation which we solved). It is good at this stage to test a few values like f(0) and f(60) to reinforce students' confidence in the function. We can now use the CAS command solve(f($x$) = 48,$x$) and find the same answer as we found using the traditional method. This technique will become very powerful as modelling with more sophisticated functions is found to be necessary.

There is considerable debate on the merits of computer algebra systems with keen advocates promoting their use against a concern for the clear requirement to effect substantial change in assessment systems predicated on routine solutions which CAS can perform for the user. (e.g., Bohm et al., 2004) In this context the CAS is being used to support and sustain the mathematical narrative. Discussing possible curriculum change in Australia, Driver suggests that CAS

> can be used to "do the messy algebra". By allowing a student to focus on the selection of a problem solving strategy or appropriate procedure rather than the application of the strategy or procedure, and student can develop their higher-order thinking skills. (Driver, 2008)

This is our purpose here. Beyond the linear case, the graphical transposition and equation solving would be difficult and would certainly get in the way of the narrative flow. Even the linear case requires effective routine facility, which if not secure will change the focus of the narrative. It does nonetheless engage the student with the need for this facility and more clearly motivate its development at a later point.

Returning to the narrative, we now have function which fits our data, so we can test its predictive capabilities. At the end of the experiment, students estimated the temperature after longer periods. In discussion, the basis on which these estimates were made changed from the short term mechanism of the linear decrease. After 30 minutes and certainly after 120 minutes most students are suspecting that the rate of decrease will have slowed. Some students suspect that after 24 hours the pizza will only have reached room temperature. So, we can test these in the function. Student's commit their expectation of the outcome to paper, first. Neatly, the CAS can deal with an input like f(24*60*60) to test the 24-hour figure. The outcomes for 30 minutes, 120 minutes and 24 hours respectively, provide an increasing surprise and realization that something is wrong. Going back to the graph and extending the horizontal axis progressively provides a visual confirmation. Students are able to interpret the graph now that they have identified the relationship between the downward graph and the cooling pizza. They are, of course, very well aware that, left to their own devices overnight, pizzas do not continue cooling, freezing and ultimately going below absolute zero! So, they are well oriented to finding a function that fits the data, but does not continue to decrease in this way.

Lesh et al. developed modelling activities for a range of groups from middle school students to graduate students, they found that,

> Few students who worked on this version went through more than half of a modelling cycle; and, almost none persevered to the point where they could make even an educated guess regarding predicted gains. After producing 'first-draft answers', these students did not feel any need to produce second- or third-draft answers. (Lesh et al., 2008)

They went on to develop their activities, but explicitly to present a second activity requiring a different analysis. The pizza scenario has the advantage that the first (linear) model is overwhelmingly favourite (with student and teacher participants) and the critique of this model is clearly grounded in participants existing knowledge of the scenario. However, the insight gained by reflecting on the basis for the estimates sets up the natural concern that the linear model isn't quite right. Hence, the second (and third) iteration appears as necessary development.

Hence, students are now free to explore different functions. The key feature they have seen is the ability to control the shape and position of the graph by varying the coefficients in the function.

TI-*n*spire has a useful facility to aid this exploration. When a base function such as $f(x) = x^2$ is drawn, its graph can then be dragged to different positions and then flexed to change its shape. The software shows the function in its completed square form, which show clearly how the graph has been transposed. An added bonus here is developing the recognition that different forms of an equation are powerful in different ways, the completed square form is often only seen as a long-winded way of solving a quadratic equation.The quadratic function $f(x) = x^2$ is the next most common function for students to meet after linear ones. Students are often aware of its existence and may have met the shape of its graph. Once in control of the coefficients, students find a quadratic which accurately fits the data. They can then use CAS tools from before to solve the equation $f(x) = 48$ and test the accuracy over the longer term. It becomes clear with this analysis that the function falls down in the longer term because it seems to suggest that pizzas will cool to a minimum and then start heating up again, becoming very hot indeed by the following day. Once again this does not accord with the students' common sense notion of how pizzas actually behave. Now, they have a very strongly formed mental image of the shape

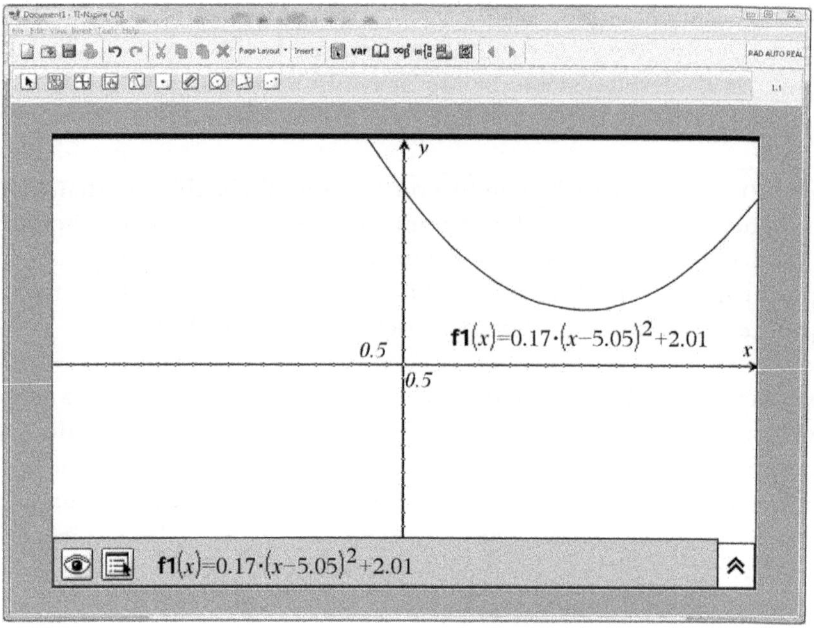


of the graph of the function they are looking for. In our lessons we restricted exploration to a palette of possibilities consisting of linear, quadratic, reciprocal and exponential function. However, the software can cope with other interesting functions, which we have tried out in teacher sessions, notably, piecewise linear functions. These naturally accord with the descriptions of the variation students suggested that is, a certain rate of decrease over a certain range, followed by a different rate of decrease over the next part of the range. With a final linear function of f($x$) = room temperature after a certain period, this can be an excellent model. Clearly, a well constructed exponential will also provide an excellent model.

## Conclusion

By this stage it is clear that we have far exceeded our requirement to find the time for the pizza to cool to 48°. Moreover we have developed our skills in finding and evaluating models for the cooling function. There is now a high degree of confidence that we can find the cooling time effectively in different situations. The experiment can now be repeated with different types of packaging. The consultants are ready to report to the pizza shop owner. With the added calculation of the distance possible given a known average speed of the delivery vehicle, the problem becomes one which geographers and business studies specialists may be better able to relate to.

I have provided a detailed description of a classroom narrative. The structuring of the narrative is built around a mathematical modelling activity, which is itself couched in a realistic but fictitious scenario. The structure is carefully designed to continually provide the rationale for further development of the theory. There are two principle mathematical outcomes: (1) an engagement with the process of mathematical modelling per se and (2) functions and their graphs.

## References

Ainsworth, S. and VanLabeke, N. (2004). 'Multiple forms of dynamic representation'. *Learning and Instruction* 14 (3) (June), 241–55.

Bohm, J., Bőjm, J., Forbes, I., Herweyers, G., Hugelshofer, R., Schomaker, G. (2004). *The Case for CAS* www.t3ww.org/cas/index.html (accessed on 6 February 2011).

Dowling, P. (1998). *The Sociology of Mathematics Education – Mathematical Myths/Pedagogic Texts.* London: Falmer Press.

Driver, D. (2008). 'Pedagogical Use of a CAS'. In '*Enhancing Understanding and Constructing Knowledge in Mathematics with Technology*'. Proceedings of ATCM, 2008. http://atcm.mathandtech.org/EP2008/papers_full/2412008_15273.pdf (accessed on 6 February 2011)

Lesh, R., Middleton, J., Caylor, E., Gupta, S. (2008). 'A science need: Designing tasks to engage students in modeling complex data'. *Educational Studies in Mathematics* 68 (2) (13 June), 113–30.

Roschelle, J., Tatar, D. , Shechtman, N., Hegedus, S., Hopkins, B., Knudsen, J., Stroter, A. (2007). *Can a Technology-enhanced Curriculum Improve Student Learning of Important Mathematics?* (SimCalc Technical Report 1). Menlo Park, CA: SRI International. http://ctl.sri.com/publications/displayPublication.jsp?ID=623 (accessed on 6 February 2011).

Teachers' TV (2006) see: www.teachers.tv/videos/ks3-ks4-maths-new-maths-technology-in-the-classroom (accessed on 6 February 2011).

# Part Six

# ICT Supporting Cross-curricular Work with Mathematics

In this section we look at some classroom (and outdoor) examples of how Information and Communications Technology (ICT) can act as a catalyst for cross-curricular work involving mathematics. The examples are selective and there are also plenty of opportunities for using for example, data-handling software for enquiries in geography and history, geometry software for exploring linkages in design technology, and so on. While cross-curricular themed project work is relatively commonplace in primary schools it is much rare in secondary schools. However the advent of the new national curriculum in England coupled with the STEM strategy for schools is providing schools with increased opportunities to develop enhancement and enrichment opportunities within subjects, and time for regular periods of 'off curriculum' time when cross-curricular activities can take place. In Part 4 we saw some examples of how new digital technologies can, if used in an appropriate way, help bring together the technologies usually associated with: (1) students' own private pursuits, (2) the way technologists, scientists, mathematicians, and so on, solve real problems and (3) the content and approaches of the school mathematics, and related curriculum areas. The contributions to this chapter show a number of innovative approaches now being used in this synthesis of use of ICT.

Chapter 18

# Using Video Analysis to Develop Modelling Skills in Physics

Steve Hearn
*Charterhouse School, Godalming, Surrey, UK*

## Introduction

In this chapter I present three techniques I use with my students to develop their modelling skills in physics lessons each using video capture as the main means of collecting data.

## Maths on Video

Imagine a powerful modelling tool. This tool can solve equations that are an attempt to describe the behaviour of the world, a model. Not only that, this tool can allow variables in the model to control the behaviour of objects on the screen. Particles and vectors, balls and arrows, become animated and move, showing the student what the mathematics is saying so elegantly. In addition video, photos and data captured from sensors can be examined on the screen using these animated objects. The effectiveness of the model at describing nature can be evaluated. Such a tool exists, *Modellus.*

*Modellus* (a popular modelling environment fore physics) facilitates the animation of mathematical models. This might involve just plotting variables. But most powerfully the teacher and student can make the output from model control the behaviour of objects on a screen. These objects in turn can be an attempt to represent the behaviour of the real world as indicated by a video running in the animation window.

The screen shot in Fig. 18.1 shows the mathematical model for undamped simple harmonic motion (SHM) as a differential equation as well as parameter choices and initial values. These are chosen to match the experimental conditions.

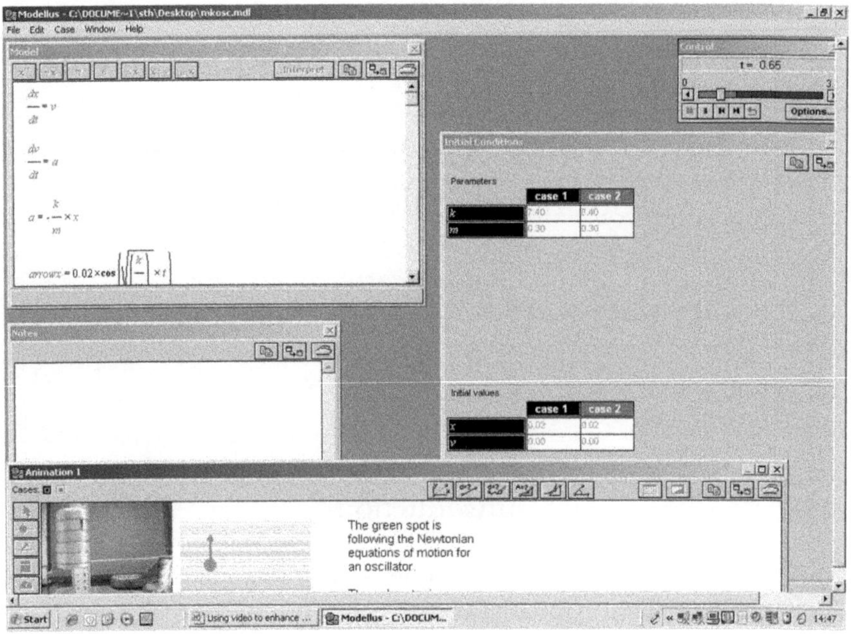


**FIGURE 18.1** Screen Shot of *Modellus* Used to Model an Oscillation

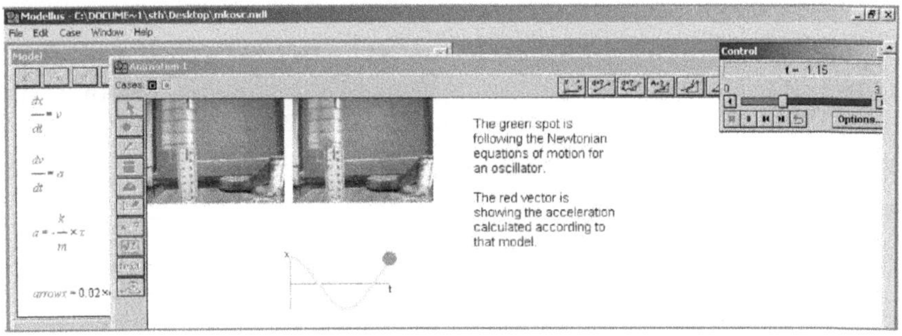


**FIGURE 18.2** Screen Shot of *Modellus* Used to Animate a Model

Figures 18.2, 18.3 and 18.4 show three further screen shots taken as the model plays in time. The equations can be seen in the model window. A mass on a spring system is being modelled. Two copies of this video have been linked to the animated objects, a vector and a particle. The vector and particle can be seen superimposed on one copy of the video.

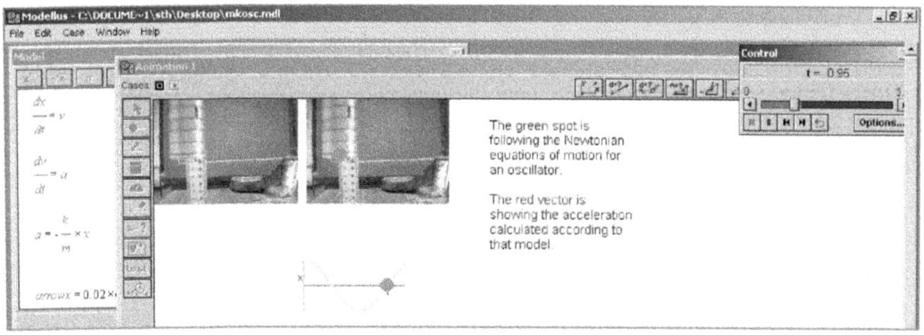


**FIGURE 18.3** The Animation at a Time When the Displacement is Zero

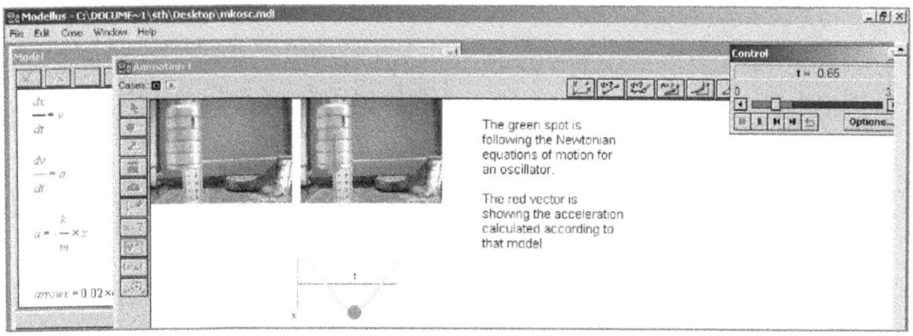


**FIGURE 18.4** The Animation at the Lowest Point of the Mass

In the animation window a point is tracking the motion of the mass and an arrow represents the instantaneous acceleration of that mass. The spring constant is denoted by *k* and the mass by *m*.

The students made a 25-frame per second video of the experiment using a standard webcam.

In Figure 18.2 the mass is reaching its maximum height. The graph of its displacement versus time, as generated by the model, is plotted. The students can watch the plot, the video and the acceleration vector simultaneously as they change on the screen.

As the mass approaches the equilibrium position (Figure 18.3) they see the acceleration vector disappear. This should prompt them to pose some questions and seek some explanations.

As the mass reaches it maximum displacement below equilibrium they observe the vector change direction.

Of course other plots can be made and additional vectors added to the animation. (It is important as an aid to understanding not to clutter up the screen with arrows and plots.)

By running this model, making alterations to the output presentation and examining the video and graphical output together, students can effectively study the mathematics of SHM.

In some cases students may have written the model, captured the video and measured *k* so as to test their mathematical representation of this motion. This is a powerful teaching experience, if a little demanding, for most sixth form pupils i.e. those aged 16–19.

For many students the video gives them a feel for the important aspects of the motion being described by the model and having the video there while they study technical output allows their understanding of the motion to develop and deepen.

Such a model can be used to examine students' understanding by supplying questions to be answered as they run the animation.

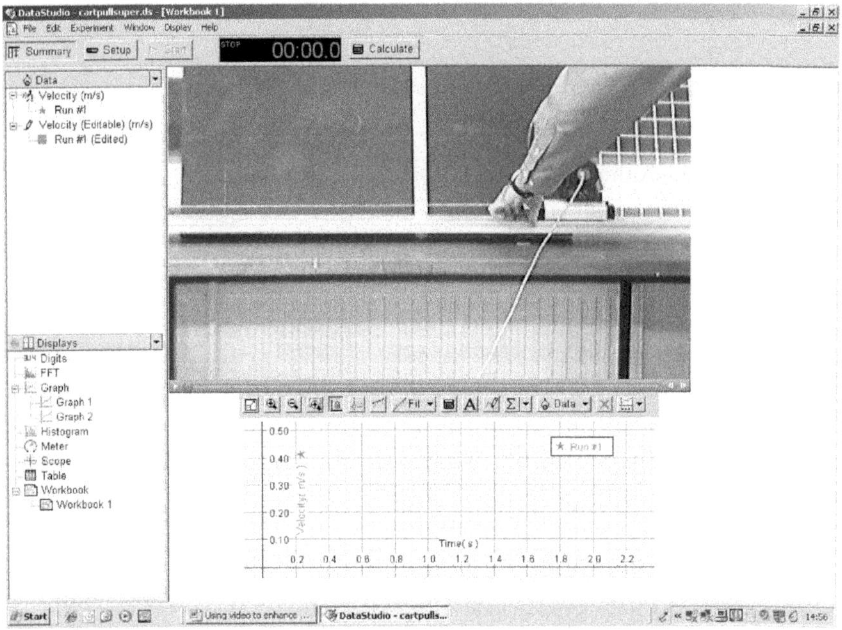


**FIGURE 18.5** Showing Video Data for an Experiment with Data Logged with a Motion Sensor

## Synchronizing Video with Data Loggers

Most schools make use of some data loggers in science teaching. Some use webcams with video pointing software. Few have explored the powerful combination available when video capture and data logging are combined.

I will describe a simple example to illustrate the technique and highlight the role it plays in the modelling process. In this experiment from mechanics, a trolley is released to roll smoothly down an inclined ramp pulled by a string from a mass falling vertically.

A motion sensor has captured velocity time data of the cart being pulled by the falling mass on a string. A video was made of the experiment and synchronized to the data. This is a simple process carried out within the data logging software, Figure 18.5.

When the video runs we see the graph of the synchronized data displayed simultaneously on the screen, Figure 18.6. Each video frame relates to a data point.

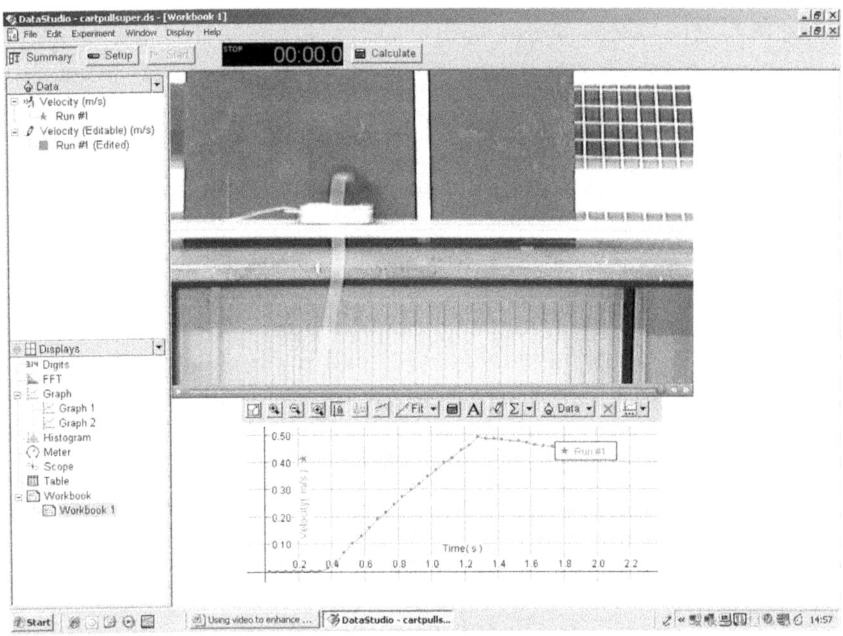


**FIGURE 18.6** The Video Data Shown with the Graphical Display of the Logged Data

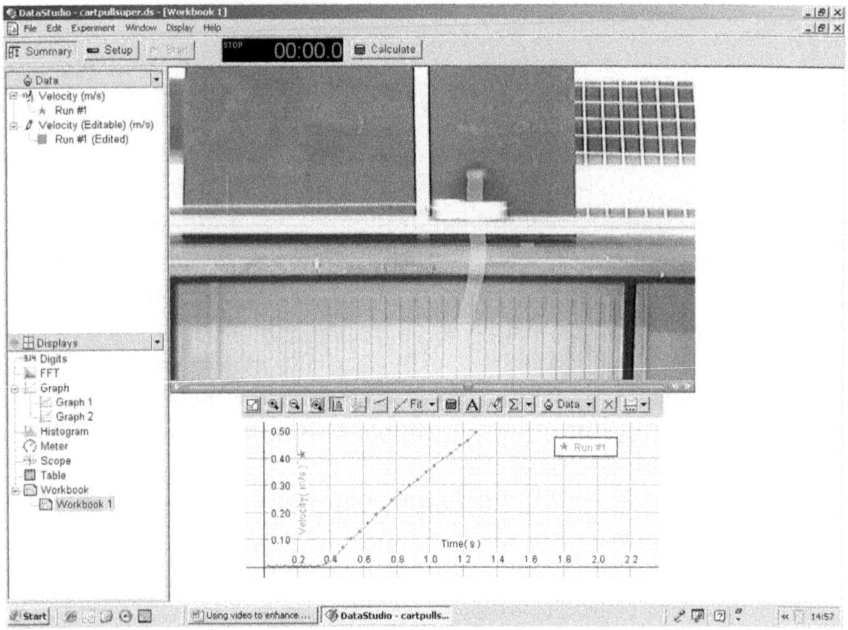


**FIGURE 18.7** The Video is Stopped Just at a Critical Point in the Experiment

Here we stop the video at a crucial point in the film. The string is just about to go slack, Figure 18.7.

The string has gone slack as the mass has hit the floor, Figure 18.8. The velocity time graph has changed drastically.

The velocity time graph has all the physical features of the video encoded. I challenge the students to draw free body diagrams for various stages of the video and then relate them to the graphical representation.

To do this successfully the students need to turn the easily digestible video into a cartoon of free body diagrams where friction suddenly appears to be significant and tension comes and goes. The more I experiment with this technique in my own lessons the more clearly I see its potential as a tool to enhance students' understanding of the processes involved and the way it is represented mathematically.

## Video Pointing

Open Source video pointing software for Physics has been developed in the USA. I use *Tracker* which is available from: www.cabrillo.edu/~dbrown/

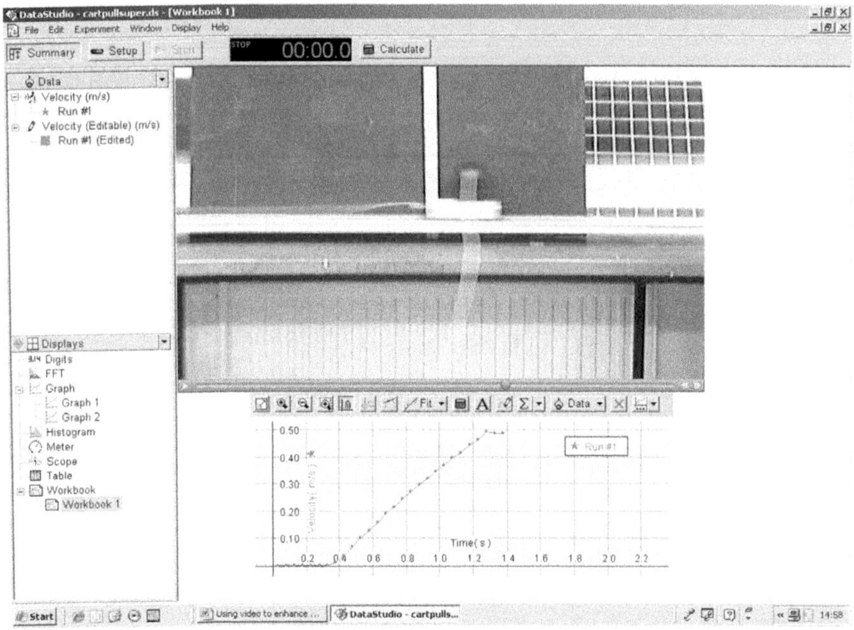

**FIGURE 18.8** The Video Stopped When the String is Slack

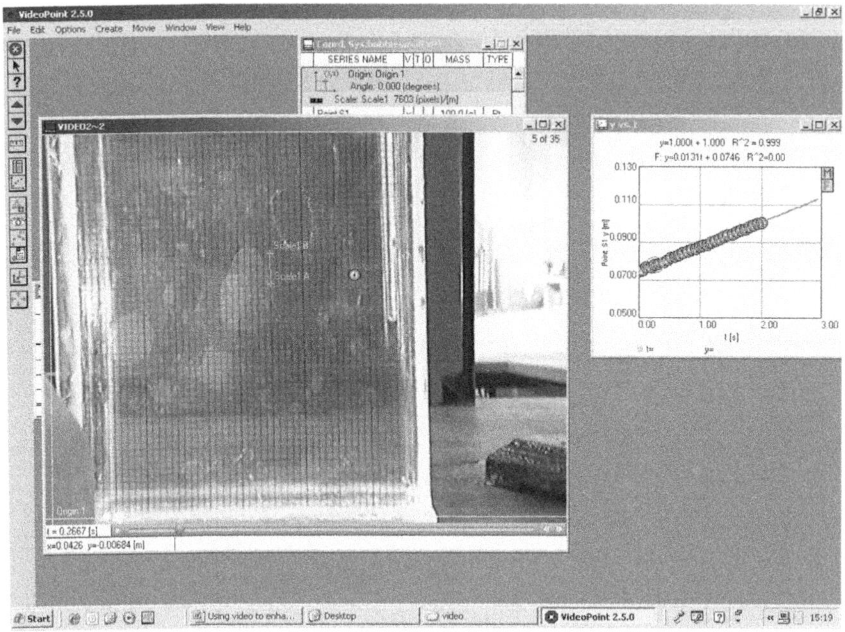

**FIGURE 18.9** A Still from a Video Clip of the Motion of an Air Bubble

tracker/. Used with a digital camera or webcam you have a powerful motion analysis and modelling tool.

Figures 18.9 shows a screen shot of a frame taken from a video clip of the motion of an air bubble. The software, which keeps track of the pixels you point to as they move, is being used to explore the question, 'Is the bubble at terminal velocity?'

## Conclusion

The data produced are good enough to allow us to study models of drag which attempt to account for the relationship between terminal velocity and size. The video technique allows good data for the size and velocity to be captured for a full analysis of drag models. Comparing this technique with the more common approach using just using a ruler and stop watch shows the students just how powerful video analysis can be. Of course this something they already know from watching sports.

The advent of reasonably cheap high quality digital cameras with the ability to record short video clips at very high speeds (up to 1000 frames per second), such as the Casio Exilim range, opens up whole new areas are opening up for video pointing analysis.

Chapter 19

# Bloodhound SSC: A Vehicle for STEM

Ian Galloway
*Science Learning Centre SE, University of Southampton, UK*

## Introduction

The very title Bloodhound invites the question, why such a name and what does SSC mean? Bloodhound was launched in October 2008 at the Science Museum by science minister, Lord Drayson. The project is to build a supersonic car (SSC) which can be driven to speeds in excess of 1000 mph (miles per hour). Designed by Ron Ayers who was involved with the Bloodhound missile, it is headed by Richard Noble and has aspirations to have a mini Apollo Effect. President Obama (2009) in his address to the National Academy of Sciences makes reference to the power of the Apollo Effect.

> The average age in NASA's mission control during the Apollo 17 mission was just 26. I know that young people today are ready to tackle the grand challenges of this century

The primary aim of the project is to help stimulate young people to consider STEM (science, technology engineering and mathematics) subjects in their career plans.

At the time of writing, more than 1000 schools have declared their interest through registering on the website and downloading materials. The principal cross curricular teaching tool is the design (technology), construction (engineering) launch (science) and evaluation (mathematics) of a mini Bloodhound car. A CPD (continuing professional development) programme has been run through the Science Learning Centre network in which teachers can learn how to use model rocket engines safely. A number of schools are using this idea to hold collapsed days at the end of the school year. There is no doubt that launching a rocket

propelled car which can reach speeds of 100 mph generates considerable excitement. What we do not know at this stage is whether or not the excitement will be sufficient to sustain an interest in STEM beyond school.

It is important to keep in mind the real engineering project which is supporting the educational programme. Two universities, Swansea and the West of England are providing the engineering teams to build and test the car. The design of the car is an amazing mathematical feat in itself, requiring 18 months of CFD (computational fluid dynamics) to produce a shape which will 'defeat' the enormous air resistance at supersonic speeds. The rocket is being specially designed and built for the car. The jet engine is from a Eurojet fighter plane and presents considerable unforeseen technological challenge as the electronics do not like being at ground level! The wheels will be made from solid titanium, but nobody has ever cast such a large titanium disc before. Safety is of paramount importance as this project is being followed by schoolchildren. The driver has to sit just beneath air-intake of the jet engine, so no ejection seat! Timing of the various operations will be critical to ensure that Bloodhound SSC reaches top speed over the measured mile in the very middle of the run. This is important as the car must make a return run over the same track within 1 hour of starting and must therefore be in the right place at the end of the first leg as the measured mile remains in the same location!

Four resources with an evident Information and Communications Technology (ICT) and mathematics link will be outlined here: the use of Google Earth, measuring the speed of sound, data logging and the slow motion tool in Windows Media Player. Mathematics will be found in all the resources but these four will exemplify the cross curricular nature of the project.

## The Hunt for the Desert (Geography, ICT and Mathematics)

With the given technology the car will need about 5 miles (8 km) to reach 1000 mph and 5 miles to slow down. With a mile at each for safety the team will therefore need a flat desert of the right hardness some 12 miles long. The considerable task of locating suitable areas was undertaken by the Geography Department of Swansea University who used algorithms and Google Earth to hunt systematically all over the globe for candidate deserts.

The exercise in the education resources makes use of the *Google* ruler on the tool bar to measure distances at any scale. Students have to locate Black Rock, Nevada (scene of earlier land speed record runs) and determine whether or not the site is long enough for Bloodhound. The primary school version gives the coordinates of Black Rock as it is not so easy to locate the target. The task reinforces the use of coordinates and additionally the altitude is given so that scales at different heights can be calculated. Does doubling the height halve the scale?

## The Speed of Sound (Mathematics, ICT and Science)

Being a supersonic car makes determination of the speed of sound important for determination of the Mach number (ratio of the vehicle speed to the speed of sound at that location). The speed of sound depends mainly on the air temperature and its determination is relatively easy using free software such as *Audacity*.

Downloading *Audacity* is straightforward. Old earphones which can be pulled about a metre apart are used as microphones and plugged into the microphone socket on the laptop or PC. By separating the earphones and clapping your hands near one of them at a point co-linear with the two phones the time difference for the pulse to travel from one phone to the other can be measured. The screen shot in Figure 19.1 shows the time difference from the peak of the first wave in the pulse to the peak of the same first wave as it arrives at the second phone. Times are in microseconds and this requires the student to handle small numbers when working out the speed. The time is returned in a small window at the bottom of the screen but can be checked using the scale at the top. The process is good enough to show different speeds at different temperatures and a model relating sound speed to temperature can be constructed.

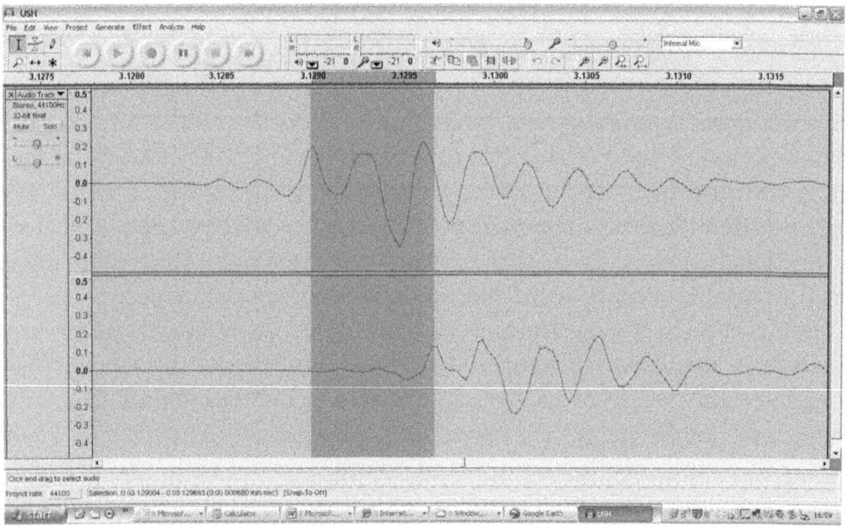


**FIGURE 19.1** Typical Result for Speed of Sound Measurement Using *Audacity*

## Impulse and Momentum, Data Logging (Mathematics, ICT, Technology and Science)

Having built a car, students could test the engine to predict the maximum speed. This can be done in the classroom but teachers may prefer to go outside for safety! The firing of the engine could be a science task and students could take their data to maths class to make their predictions but in any case the exercise is a mathematical one wherever the analysis is carried out! The previous experiment to measure the speed of sound can also be described as data logging but this one uses commercial apparatus which all science departments will have. A force sensor is set up with the car's nose against it. When logging is commenced the rocket engine is fired and the thrust data collected, Figure 19.2.

In the example in Figure 19.3 the area between the graph and the time axis has been highlighted and the software returns the impulse in Ns (Newton seconds). Since impulse equals change in momentum the maximum speed can be calculated if the car's mass is known. The acceleration can also be calculated so that the length of the track needed to reach top speed can be estimated. The second blip on the graph is the parachute charge being ejected from the engine.

This exercise reinforces use of equations and develops graphical interpretation as well as being a technological process yielding scientific data.

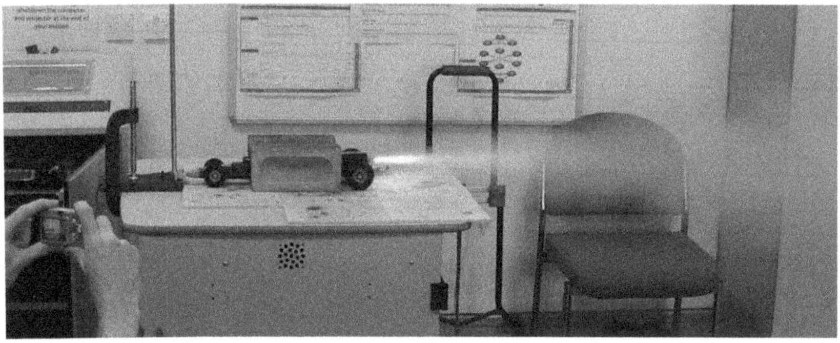

**FIGURE 19.2** Test Launch in a Classroom, Smoke Detectors Off!

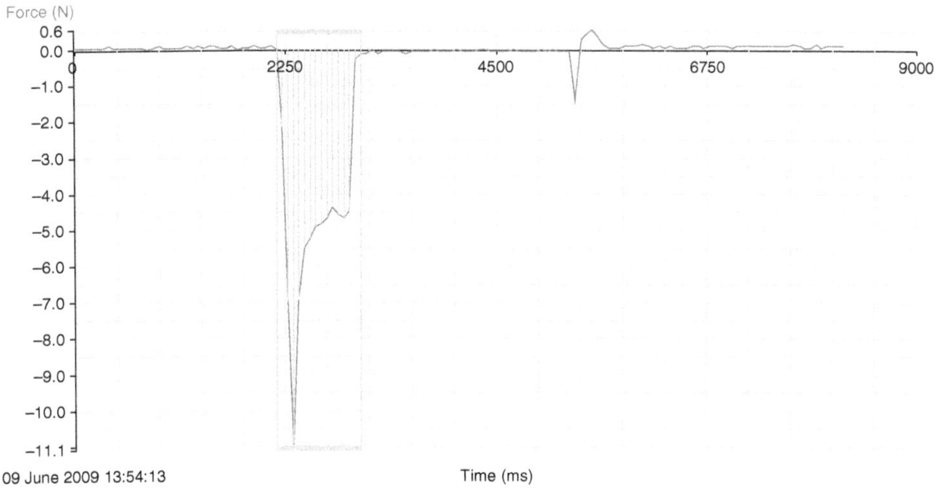


**FIGURE 19.3** Typical Results, Force Against Time
(note the inversion due to the sensor being pushed rather than pulled)

The Bloodhound engineers have had to go through the same process for the real car.

## The Motion (Mathematics, ICT, Technology and Science)

Reading the graph to understand what is happening during the motion is in itself a challenging exercise.

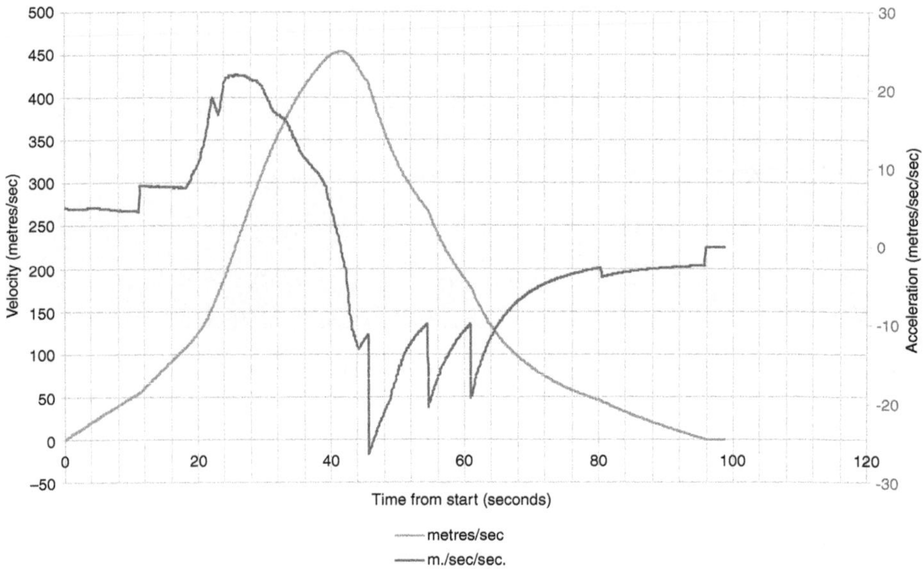


**FIGURE 19.4** Profile of the Motion of Bloodhound SSC

The two graphs shown in Figure 19.4 are the velocity (smooth) and acceleration curves. There is a great deal that students can identify with the events taking place during the run such as the opening of parachutes and ignition of the rocket. The real cross-curricular approach comes when students launch their model car and attempt through various means to analyse the motion and compare it to their predictions.

Figure 19.5 shows how the car was filmed and played back with *Windows Media Player*. Chalk marks have been made on the ground and the play speed settings selected (right click in the control bar, then select *view, enhancements* and *play speed settings*). This allows the video to be advanced a frame at a time and providing the frame speed is known the (average) speed can be measured at different points along the track.

Some schools have organized competitions around the real engineering challenge. It turns out that using a C sized model rocket engine will require a 50 m track, and even then a substantial amount of bubble wrap will be needed at the end of the track for the car to be brought to a halt! The real track will be about 10 miles so the measured mile is about a tenth of the whole distance. In the model world a 5 m section of track will therefore correspond to the measured mile. The challenge is to place the beginning and end markers of the 5 m section and film over that part to measure the speed reached.

**FIGURE 19.5** A Still from a Film with Chalk Marks

## Conclusion

There are other resources on the Bloodhound website which attempt to integrate the STEM subjects. The view taken within the Bloodhound team is that engineering is a STEM profession. Any engineering project is a combination of mathematics, science and technology and represents a real STEM activity, Bloodhound SSC being no exception. There are links to earth science, history and geography so that a true sense of a real world task needing the application of a large range of skills emerges. There is no escape from the all pervading presence of mathematics and the idea that progress can only be made with its application. The mathematical applications need not be complex but they are ubiquitous even where not explicit. Time will tell whether or not the desired Apollo effect is achieved but the early part of the story shows that children are excited by this real world engineering project to re-take the World Land Speed Record.

## References

All websites accessed on 6 February 2011.

Audacity, http://audacity.sourceforge.net/
Bloodhound, www.bloodhoundssc.com
Google Earth, http://earth.google.co.uk/
Obama, B., (2009). www.washington.edu/externalaffairs/fedrelations/president-obama-makes-remarks-to-national-academy-of-sciences/

Chapter 20

# Modelling Action in Sports and Leisure

Matt Pauling
*National Development Manager, Youth Sport Trust*
Adrian Oldknow
*Emeritus Professor, University of Chichester, UK*

## Introduction

Hampshire LA ran a 2-year pilot project in 2006–2008 with 5 secondary schools each identifying a maths, science and D&T teacher to work with other colleagues using Information and Communications Technology (ICT) to stimulate cross-curricular work. The project, directed by Ron Taylor, was supported by the Microsoft/TDA `Partners in Learning' initiative. Henry Cort School is a Sports College which developed ICT based approaches for links with maths, science and D&T in sport. These have been carried forward into more specifically STEM based R&D and CPD activities in the region. Some of the Henry Cort work is featured on Teachers' TV: www.teachers.tv/videos/quadratic-equations. It was a case study for the Youth Sports Trust's '*Raising Your Game*': www.ncetm.org.uk/files/360544/Raising+Your+Game+coresubj_070307.pdf.

## Types of Activity

The main activities involved throwing, kicking, hitting, jumping, firing, and so on. Technically these are forms of **projectile motion**, which was the basis of the study of dynamics (forces, acceleration, velocity etc.) which led to the discoveries and theories of Galileo and Newton. A simple example is throwing a small hard ball at an angle. In order to find data about its motion we have three main techniques, each of which involves ICT:

- Measurement and modelling

- Data capture from sensors
- Data capture from video clips.

## Measurement and Modelling

Measurements can be made by estimating, or with simple techniques like counting and pacing, or using instruments like protractors, tape measures and stop watches. For many sporting events, such as sprinting or track cycling, these are not accurate enough and computer based measuring systems are used. Some games require judgment for example, to decide whether or a ball, or batsman, was in or out. The human line judge, referee and umpire is increasingly being supported by ICT based systems, such as tennis's Hawkeye system.

For a projectile we can measure the launch angle A°, horizontal range R m. and flight time F s. Assuming ideal conditions, such as no air resistance, spin or wind, and that the launch was from ground level, then the science theory states that the only force acting on the object after launch is that of gravitational attraction – so the acceleration downwards is $g$ $ms^{-2}$ and horizontally is 0 $ms^{-2}$. From these, together with the ideas of speed as distance divided by time, and acceleration as change in speed divided by time, we can derive formulae for the motion such as $H = R/F$, $V = H \tan A$, $g = -2V/F$ and also find the position $(x,y)$ at time $t$ as $x = Ht$, $y = Vt - \frac{1}{2} gt^2$ . This use of algebra to find features of the motion based on scientific principles is called **mathematical modelling**. The algebraic formulae can be embedded in a spreadsheet and so be used to create a **simulation** of the ball's flight. Using mathematical software such *Geometer's Sketchpad, Cabri II Plus, TI-Nspire, Autograph* – or graphical calculators and TI-Nspire hand-helds – graphs can be made directly from the measurements and calculations as shown in Figures 20.1 and 20.2. These models can be used to predict behaviour for different launch angles, projection speeds, and so on. The predictions can be tested practically to see how well the model matches with reality. Note that this gives an experiment from which gravitational acceleration $g$ can be measured.

## Data Capture from Sensors

A common system used in science for timing events involves light gates connected to data-capture software on a computer. Knowing the positions of

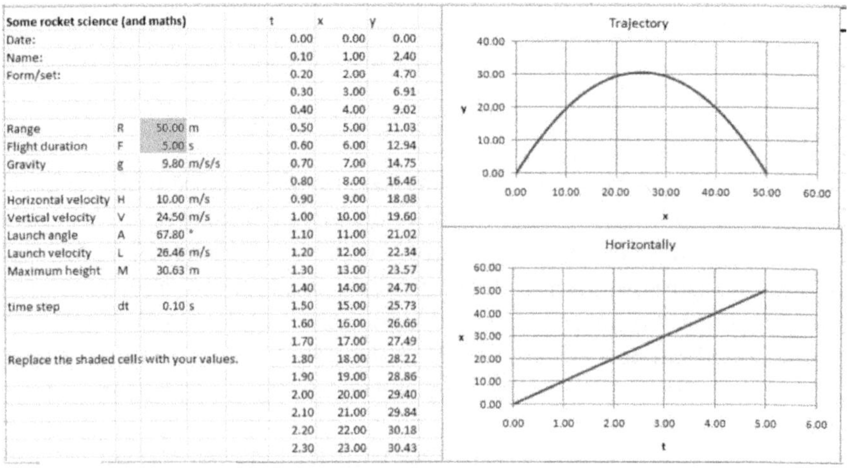


**FIGURE 20.1** An *Excel* Spreadsheet Projectile Model

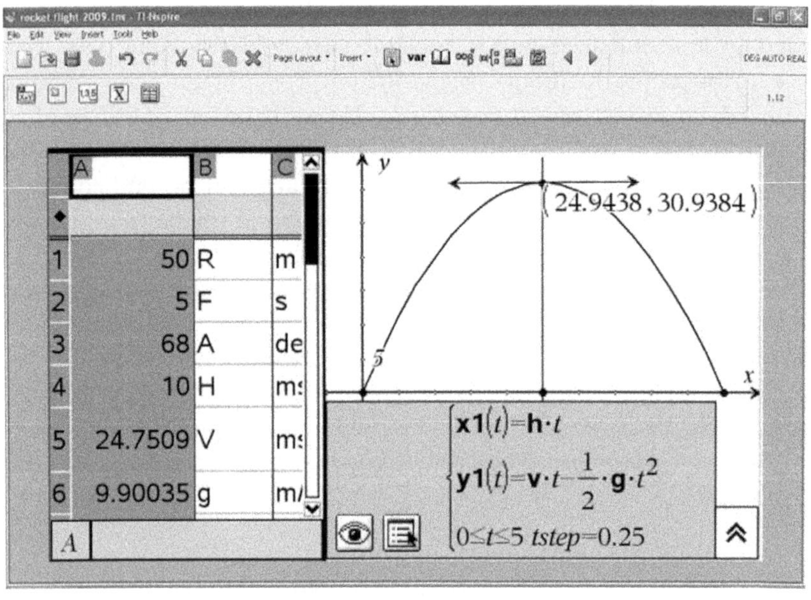


**FIGURE 20.2** A *TI-Nspire* Projectile Model

the gates, and finding when an object passes through each, we can compute its average velocity. A range-finder, or motion detector, can measure distances to a given object in motion and be used to plot its distance-time graph, while also computing numerical approximations for velocity and acceleration. Such a sensor (CBR2) is shown in mathematics lessons on Teachers TV programmes

- KS3/4 New Maths Technology: www.teachers.tv/video/154
- Resource review: secondary maths 2: www.teachers.tv/video/4872.

Video clips of its use, and a lesson plan, can be downloaded from the NCETM website – on which you will need to register first: www.ncetm.org.uk/Default.aspx?page=14&module=com&mode=102&comcid=241&comf=40093&comu=0 .

A more sophisticated approach uses one or more accelerometers attached to the object in motion – in this case probably a human runner, jumper, cyclist, and so on. Speed data can also be captured with a radar or laser speed gun. Once a data set has been captured it can easily be transferred into a spreadsheet, or software for data analysis such as *Fathom* or *TI-Nspire*. *TI-Nspire* hand-helds and software support direct data capture from probes such as the CBR2 and low-g accelerometer. Another approach uses a wireless system, such as Vernier's Wireless Dynamic Sensor System, together with a Bluetooth receiver and laptop PC: www.inds.co.uk/education/edusyswdss.htm. Using this system strapped to an object, data can be captured and displayed in real-time. Another exciting development supported by Vernier's *LoggerPro 3* software is to synchronize graphs of captured data with video clips of the experiment. So, for example, we can compare the acceleration of a sprinter leaving the starting blocks with video of the action. Figure 20.3 shows a video of spring-mass system together with force and acceleration graphs.

Another development is the data capturing element of the Sciencescope rocket: www.sciencescope.co.uk/rocketlogger.htm. This consists of a thin plastic tube containing three accelerometers, a pressure gauge, a timer and a battery. It can be charged up from a USB port, detached, attached as the payload for a rocket, launched from an Airburst rocket kit, recovered, reconnected and the data downloaded. The output can be data in a spreadsheet format (CSV) and/or a simulated video clip in Google Earth (KMZ). The data logging unit can also be strapped to an athlete or to a suitable piece of equipment, such as a javelin, although Google Earth output will not always be possible. Another possibility is opening up based on GPS technology: www.vernier.com/probes/vgps.html. Many of these approaches

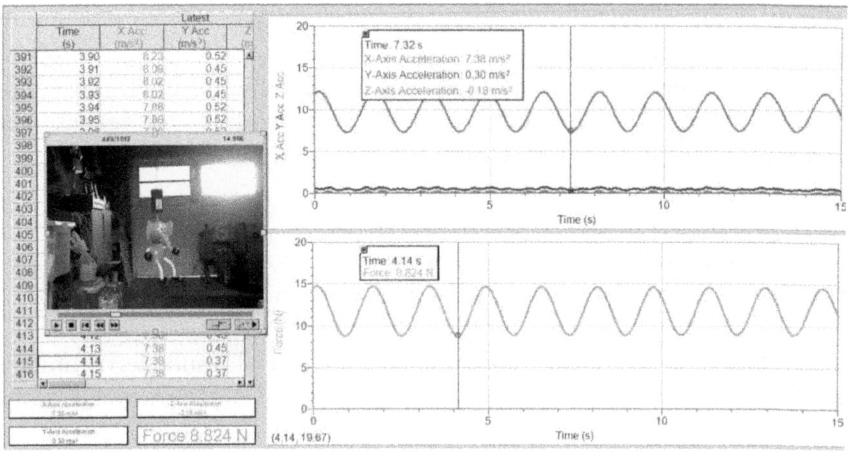


**FIGURE 20.3** Vernier's *Logger Pro 3* and WDSS

are suitable for prototyping within practical work for D&T, ICT or Electronic Systems – or by skilled teachers and technicians. We are now investigating prototyping systems such as the Italian Arduino control boards and sensors: www.arduino.cc/en/.

## Data Capture from Video Clips

Video analysis systems for use in sports have become relatively established in schools through systems such as *Dartfish, Quintic, Swinger* and *Kandle.* The use of *Dartfish* with video capture of a badminton service in Arnewood College is another DfES video case study: www.ncetm.org.uk/Default.aspx?page=14&module=com&mode=102&comcid=241&comf=40084&comu=0

Here digital video from a camcorder is downloaded to *Dartfish,* where positions of the shuttlecock are annotated on successive frames. The resulting screen image is captured to the clipboard and pasted as the background in *Geometer's Sketchpad* over which graphs are drawn and manipulated to give good mathematical models. The analysis is done in mathematics lessons, while the theory and explanations are done in science. Figure 20.4 shows a similar image captured from a free throw at basket-ball. The known height of the basket enables calibration and a variety of quadratic models can quickly be tried out. This technique of bringing the outside world into the

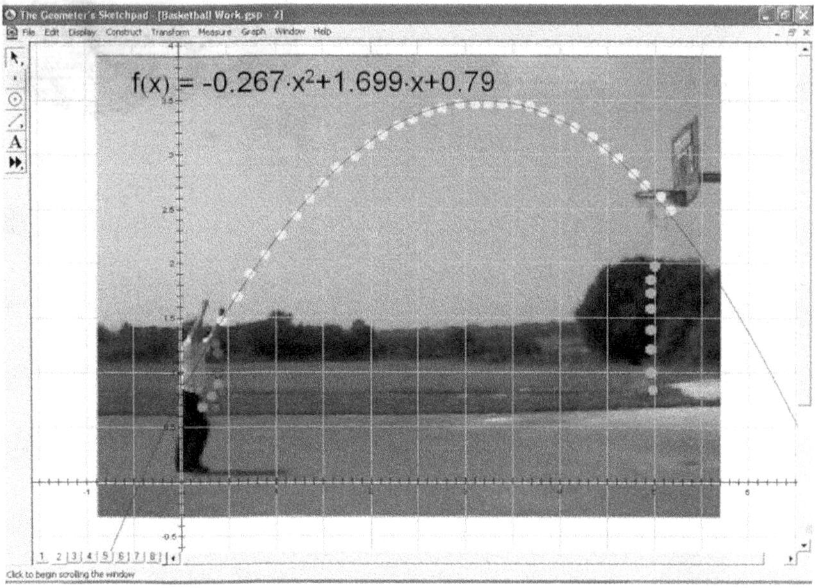


**FIGURE 20.4** Mathematical modelling with *Sketchpad*



mathematics classroom was used in a recruitment poster by the TDA in 2008 – see Figure 20.5.

The good news is that video data capture can be achieved using relatively cheap devices such as the video mode on digital cameras and/or phones or cheap video devices like Disgo and Flip cameras. There are low cost science packages such as Vernier's *Logger Pro 3*, which support video data capture, but there is also open-source free US software. Henry Cort originally used the *Vidshell* package developed by Doyle V. Davis, which you can download together with a library of video clips from: http://webphysics.ccsnh.edu/vidshell/vidshell.html. This is used with a screen shot to capture an annotated image of the position of a basketball as the background for analysis in *Sketchpad.*

As part of a QCA project 'Engaging mathematics for all learners', Teachers TV has filmed pupils from Wildern school undertaking a playground project in which they also use video analysis: www.teachers.tv/video/37909. They use *Tracker*, a free Java applet from Doug Brown: www.cabrillo.edu/~dbrown/tracker/. With *Tracker* you import a video clip, calibrate it, overlay axes and record data while tracking the position of an object. While

**FIGURE 20.5** TDA Poster 2008

the video is being annotated in one window, a table of data is generated in another, and a graph in the third. All the analysis and modelling can be done within *Tracker* itself. At Wildern a group of Year 10 students produced a DVD of their project which includes a section demonstrating how to use *Tracker*. www.ncetm.org.uk/files/362726/Wildern+clip+2.avi .

When filming motion it is important to keep the camera still, and not to track the object, nor to zoom. Normally the motion being studied only requires a few seconds of video, so access to editing software is useful. Some formats may not be readable, so video conversion software can be useful.

There is now a new range of personal digital cameras capable of high-quality video capture at speeds of 210 and 420 frames per second, and at lower quality at 1000 fps. Figure 20.6 shows the capture of the service action of Juan del Potro on his way to the final at the 2009 ATP World Tour tennis finals at the London O2 stadium last year. The video clip: 'del potro.avi' was recorded at 210 frames per second using a Casio Exilim EX-FH20 high speed camera with telephoto lens: www.casio.co.uk/products/Digital%20 Cameras/Exilim%20High%20Speed/EX-FH20BKEDA/Technical_ Specifications/.

The clip was imported into the *Tracker* software and the 'tape measure' tool was used to mark the distance between base and service lines as 5.5m. Axes were added and titled. The path of the ball was tracked as a 'Point mass' from the point of delivery by the server. The best-fit line for displacement $r$ against time $t$ is found as $r = 49.14\ t + 0.05$. This suggests that the ball is moving horizontally from right to left at a speed of nearly 50 $ms^{-1}$. This

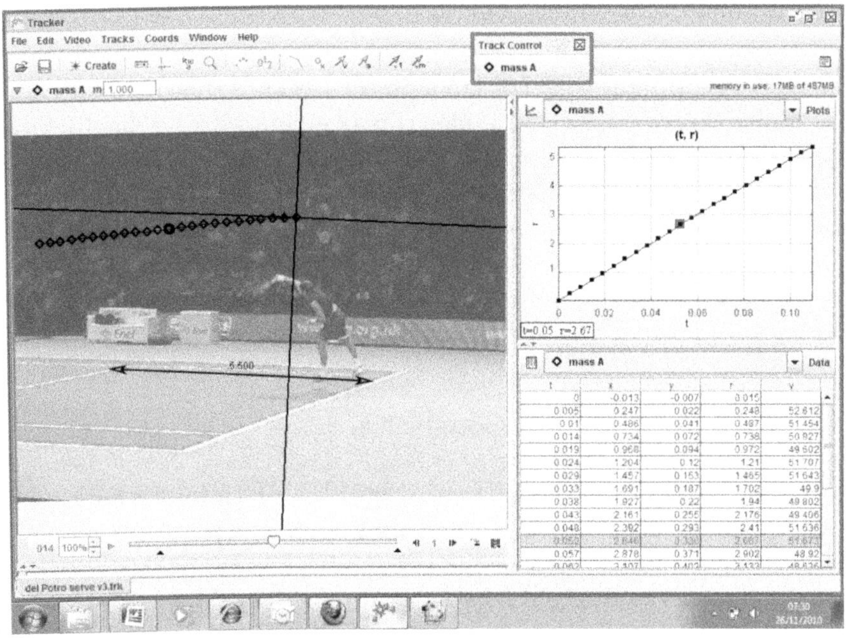


**FIGURE 20.6** Tracking the Speed of del Potro's service

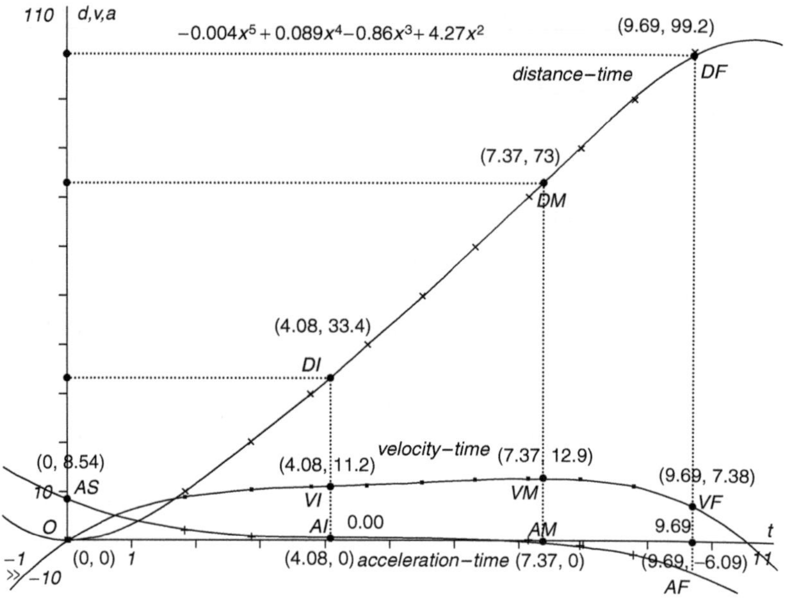


**FIGURE 20.7** TI-Nspire Analysis of Bolt's 100m Run

compares well with the speed of 115 mph (or 184 kph) shown on the court-side display. You can download a video clip together with a Tracker file from which you can estimate the speed of delivery of a ball bowled by a young cricketer from: www.ncetm.org.uk/Default.aspx?page=14&module=com&mode=102&comcid=5265&comf=45648&comfb=1&comu=0. Figure 20.7 shows a more sophisticated example of mathematical modelling applied to data from Usain Bolt's Gold Medal winning 100 m sprint at the 2008 Olympics.

## Conclusion

This article has concentrated on moving objects – which can include pupils doing sports and PE. Of course there are plenty of opportunities for ICT use to study the impact of exercise on pupils. Sensors such as heart-rate monitors, temperature probes, balances, pressure pads, force gauges, and so on can be used to measure aspects of performance. Another related field of study is biomechanics – which concentrates on the geometry of the human frame – which can be modelled using CAD and geometry software.

Another concerns the properties of new materials developed for sports clothing, such as swimwear. Yet another is concerned with the design of the apparatus used for sports such as cycling, rowing, sailing, skiing, archery, and so on. So clearly the field of PE/sports is a rich source for ICT assisted cross-curricular project work – especially for the STEM subjects.

Among other significant step forward is the introduction of a range of rugged personal sub-notebook computers running Windows XP by Intel based on their new Atom processors. These Classmate PCs are specifically designed for education within the 'one laptop per child' ethos. The current top of the range model is a tablet PC with 60 GB hard drive costing around £300.

Another new device is a GPS sensor which plugs into the USB port of a mobile PC, such as Classmate, or a mobile data logger, such as LabQuest: www.vernier.com/gps/. An alternative approach is to use a hand-held GPS navigation device, such as a Garmin eTrex: www.sciencescope.co.uk/pdf/Page_by_page_catalogue/sciencescope-p08.pdf. In either case data can be downloaded following a journey (e.g., on foot or by bike) for latitude, longitude and elevation against time for analysis for example, in *Excel,* and also for geographical display in either Google Earth or Google maps. The data captured by one or more sensors for example, heart-rate can also be displayed as a graph or using colours, and images captured en route can also be displayed.

# Part Seven

# Case Studies of Teachers Engaging with ICT

In this section teachers outline their own approaches to using Information and Communications Technology (ICT) in the classroom to support pupil learning and understanding. The issues surrounding ICT use are considered from different perspectives, for example from that of a Head of Department in a secondary school and from the point of view of Key Stage 2 practitioners. Some of the ICT used within these case studies is firmly established and recognized by practitioners as having merit in supporting understanding and learning. Newer technologies and those with little recognition as classroom resources are also explored. It is often heard commented that classroom use of ICT lags behind use in general society, it is interesting to reflect upon the reasons surrounding this issue, not only in terms of physical resourcing but also in terms of attitudes towards novel technology being employed by pioneering teachers. What is deemed as acceptable use and what might be perceived as either 'just playing' or even 'cheating'?

Part [illegible]

# Case Studies of Teachers Engaging with ICT

[illegible]

Chapter 21

# Teaching International Baccalaureate Mathematics with Technology

Jim Fensom
*United World College, SEA, Singapore*

## Introduction

IB diploma students select one subject from each of six subject groups. Three are taken at higher level (HL), the others at standard level (SL). All students must take a mathematics course. There are four different courses designed for different types of students. Mathematical studies SL is designed to build confidence and encourage an appreciation of mathematics in students not planning to continue their study further. Mathematics SL caters for students who expect to need a sound mathematical background as they prepare for future studies in subjects such as chemistry, economics, psychology and business administration. Mathematics HL caters for students expecting to include mathematics as a major component of their university studies, either in its own right or within courses such as physics, engineering and technology. Further mathematics SL is for students who have a considerable interest in the subject. Students are expected to have access to a graphic display calculator (GDC) at all times during each of these courses.

In the external assessment of these courses students also have access to a GDC, except in non-calculator papers for mathematics SL and HL, but not all questions will necessarily require the use of the GDC. In the internal assessment of mathematical studies SL students can optionally make use of technology, but in the internal assessment of mathematics SL and HL it is a requirement as the use of technology is formally assessed.

Making the GDC required in examinations rather than being simply permitted led to a new dimension in the use of technology. When it was only an optional tool, some teachers had paid little attention to teaching the skills required to use it and similarly many students made little attempt to master those skills. Once it became a requirement, teachers took the technology on board and students paid more attention to using their calculators. As a result, it became possible to embed the technology in lessons. What has been particularly rewarding is the way in which our studies students have had their interest in mathematics renewed as the calculator has given them access to areas of mathematics that had previously been inaccessible to them. Another consequence has been the growth in the use of other technologies such as graphing, dynamic geometry and data analysis software. While this had been possible before, somehow it never really took off even though the software was available and training provided.

Let us take the introduction to the differentiation of sine functions as an example of the use of technology in the teaching of the SL course. There are many approaches to doing this ranging from a rigorous analytical one to simply presenting the students with the result. This approach could utilize a GDC or any graphing software. When students in the class are comfortable with technology, there are many possibilities for introducing topics that do not need the depth required in a rigorous approach. A better understanding can be achieved than from handing students unjustified results.

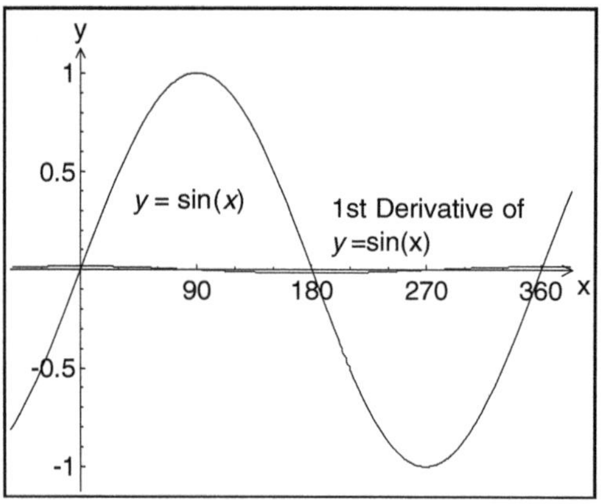


I begin by getting the students to plot a sine curve and its first derivative with their calculators (or software) in degree mode. At first sight the derivative curve is almost invisible as it runs so close to the *x*-axis. But, on closer inspection, it is clear that there is a wave and that maybe it is a periodic function with a very small amplitude.

The next stage is to experiment with the period of the sine function and notice how, although the amplitude of the original function is unchanged, that the amplitude of the derivative, clearly a sinusoidal curve, changes. This

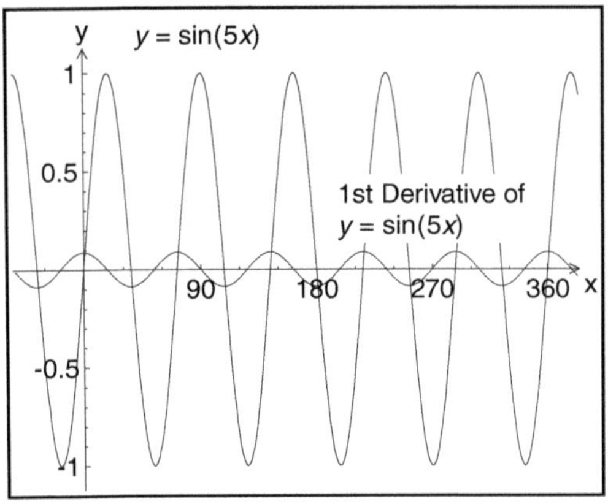


leads to the conjecture that it might be possible to find a period that makes the derivative have an amplitude of one, the same as the original sine curve.

After some trial and error and adjustment of the scale of the *x*-axis, it is discovered that a value of approximately 57 for the coefficient of x yields the appropriate amplitude. The *x*-intercepts of the modified sine curve are-approximately $\pi$ and $2\pi$ and, of course, $57 \approx (180/\pi)$. Thus the first derivative of $\sin((180/\pi)x)$ can clearly be seen to be $\cos((180/\pi)x)$ leading to the idea of radian measure.

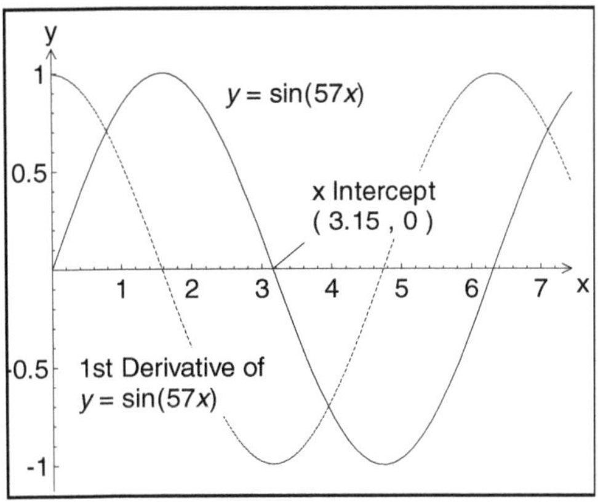


In mathematics studies SL, students undertake a project worth 20 per cent of their total mark. This could be from any area of mathematics. Access to powerful software could result in students simply adopting an approach that relied on, for example, a statistics package to do all the serious mathematical work for them. It is possible, however, to enable students to make use of the software to perform more mundane arithmetical tasks, while allowing them to interpret their results and to develop their mathematical insight.

One of my students, not one of the most mathematically able nor particularly interested in studying mathematics, was interested in football and I encouraged her to choose a topic for her project that was based on this interest. She took as her data the Premiership league table for the previous season. The question she chose to ask was, 'Which is more closely related to the final position in the league: goals for, goals against or goal difference?' She had learned about finding the least squares regression equation and the product moment correlation coefficient in class and knew how to get these with her calculator as well as how to put the data in the formulas. This example shows how the use of a GDC to enable a student to carry out a relatively sophisticated piece of mathematics can aid understanding of what that mathematics means in a very effective way.

The data was first entered into lists on the GDC. An example of what this looked like is shown.

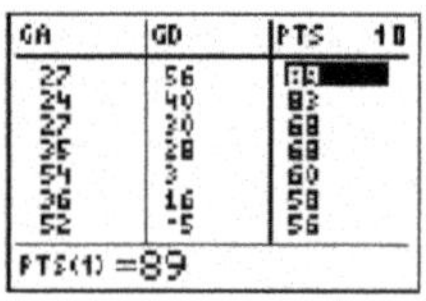


Scatter graphs of the three sets of data were drawn: goals against (GF) and points (PTS), goals against (GA) and PTS and goal difference (GD) and PTS.

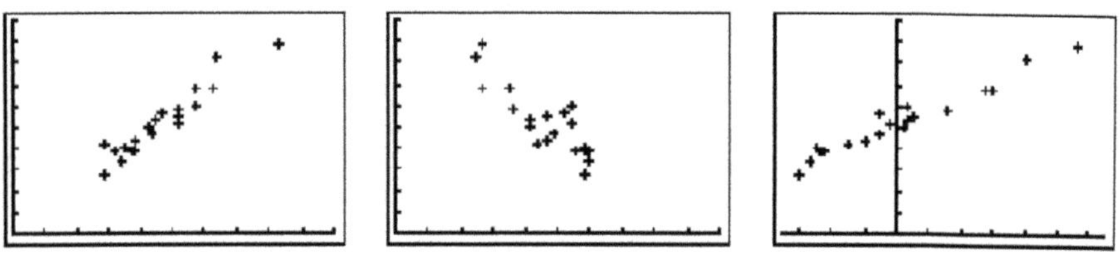

The pictures alone provided a lot of opportunity for a meaningful discussion of the strength of the correlation and whether it was positive or negative. Next the equations and coefficients were calculated and the results illustrated by graphing the regression lines superimposed over the data and conclusions were drawn.

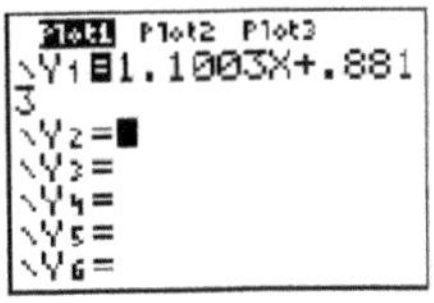


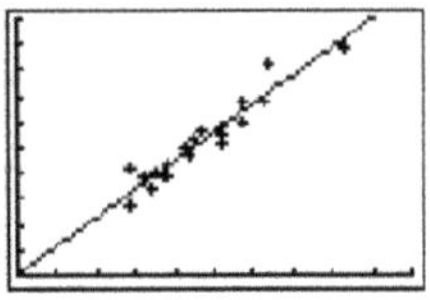

An interesting extension to the original question was to look at influence that replacing goal difference with goal average had from the 1976–1977 season in the English Football League onwards. Another list was used to calculate the goal ratio (GR) and this was plotted against PTS.

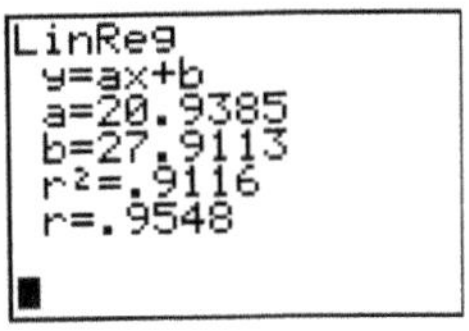


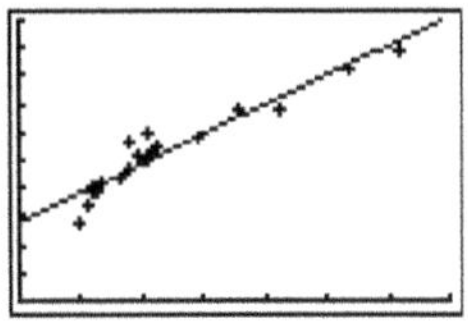

The evidence above seems to suggest that the use of goal difference was justified, certainly if only linear regression is to be considered.

Finally I would like to give an impression of the GDC skills that would be required in the examination. The following example is based on a type of question that might be found on an HL examination paper.

> On the same axes sketch the graphs of the functions, $f(x)$ and $g(x)$, where
>
> $f(x)=4-(x-1)^2$ for $-2\leq x\leq 4$
>
> $g(x)=e^{(x-2)}+2$ for $-2\leq x\leq 4$

The functions are entered into the GDC and their domains are used to determine a suitable window for sketching the two curves. The sketch is drawn from the GDC screen with axis labelling and scales added.

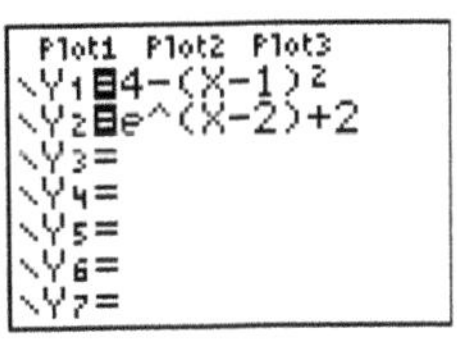


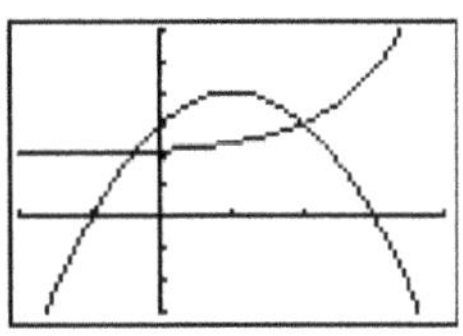

Write down the equation of any horizontal asymptotes.
State the $y$-intercept of $g(x)$.

The horizontal asymptote can be seen by inspecting the graph, together with a little knowledge of the exponential function. ($y$=2)

The $y$-intercept can then be found directly from the GDC or by substituting $x = 0$ into $g(x)$. ($e^{-2}+2 \simeq 2.14$)

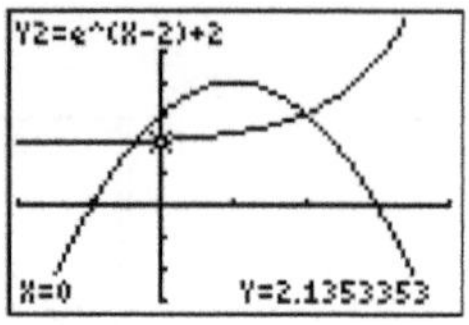


Find the values of $x$ for which $f(x) = g(x)$.

This equation can be solved by finding the points of intersection of the curves. ($x \simeq -0.381$, $x$=2 )

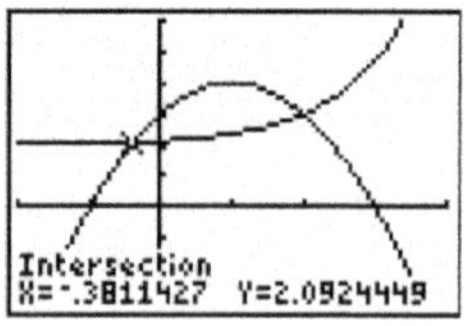


Let $A$ be the region where .
On your graph shade the region $A$.
Write down an integral that represents the area of $A$.
Evaluate this integral.

The region between the two curves to the right of the $y$-axis is shaded.
The required integral is $\int_0^2 \{(4-(x-1)^2)-(e^{(x-2)}+2)\}dx$

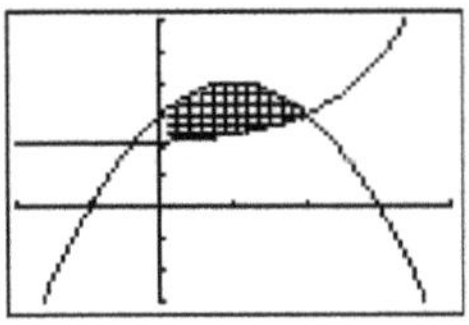

A nice way to evaluate this integral using the GDC is to plot the function $f(x) - g(x)$ and use the graphical integration option. ($A \simeq 2.47$)

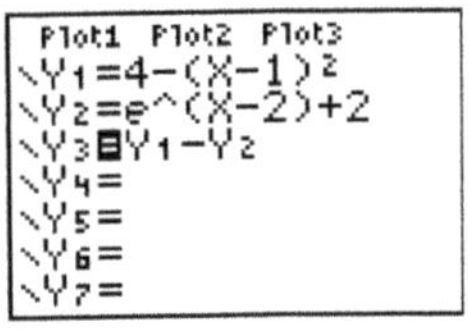


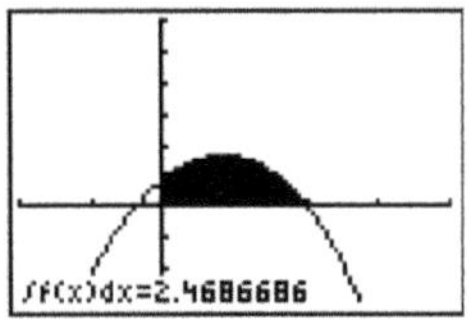


> In the region $A$ find the maximum vertical distance between $g(x)$ and $f(x)$.

Since the function $f(x) - g(x)$ has already been plotted, all that is left to do is to find its maximum value. (maximum distance $\simeq 1.66$)

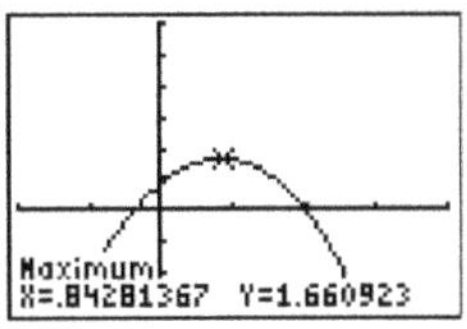


It is true to say that initially there were a number of teachers who felt that this type of problem was debasing the examination. Even among students there is still a common feeling that an 'analytical' approach is somehow more valid than simply pressing buttons on a calculator. A closer look at what is actually being examined by this question, however, reveals that there is a lot more required in order to answer it than just pressing buttons. There are other places in the paper where some of the more traditional skills can be examined, but this style of question is suited for testing conceptual understanding.

## Conclusion

I have intended through these three examples to illustrate different aspects of the way in which technology contributes to how mathematics is taught and learned in the IB. In the first, I have tried to show how technology based techniques can be used effectively to introduce topics to students who are already familiar with using mathematical software or a GDC. The second example shows how students who are not necessarily the most mathematically skilled can use relatively sophisticated techniques and develop their understanding of mathematics without becoming lost in computation. Finally, I give an example of how the examiner can test the underlying concepts of a problem and not simply the skills required to arrive at a result. Technology has been incorporated into mathematics by the IB in a way that has enriched the course and is motivating to the students. On the other hand, these positive benefits may not always be as long-lasting as we teachers expect them to be. A former student said to me recently 'I got out my TI-83 Plus for the first time since the last IB SL mathematics examination! Woah . . . this thing looks like a space shuttle dashboard!'

Chapter 22

# Why Use Technology to Teach Mathematics?

Andy Kemp
*Head of Mathematics, Taunton School, UK*

## Introduction

The debate about the use of technology in the teaching of mathematics continues on in schools up and down the country. Often it starts with conversations about the use of the humble calculator – when do you allow your students to use a calculator in class? The answer to this question says a lot about a teacher's stance both on the use of technology and also about what they consider the important elements of mathematics to be. Is the calculator another tool in your students' toolkit or is it something to be avoided wherever possible? Is using a calculator just a permissible form of cheating? Every mathematics teacher has an opinion on this question and that tends to influence their opinion on the use of technology in general.

My intention is to explore several aspects of how I use technology in my day to day teaching, exploring the impact both on me as the teacher, and also my students. I will try and outline the case for why technology in the mathematics classroom is not something to be feared but something that should be embraced for the benefit of our students. Some of the technology I will explore is inherently mathematical whereas others are more general, but all of them impact upon the way I teach mathematics.

## Tablet PCs/IWB

The use of a tablet PC or interactive whiteboard is far from unique to the mathematics classroom but it has huge benefits in the post-16 classroom when used thoughtfully and in conjunction with other technologies. For the last 5 years I have taught almost every lesson using a combination of a tablet PC and projector. I use a package called Microsoft OneNote as my whiteboard environment – It enables me to keep my notes organized by class on a collection of infinitely extendible pages. It's a bit like having one of the old roller boards but one where you can roll it back to any point in the past and can always roll it forward to a new page. The advantages of this may not be obvious at first glance, but it enables me to quickly and easily refer back to work we did at any point in the class and talk through past examples or questions when leading in to a new topic.

The other great feature of this type of approach is that it is simple for me to print off class notes for students who miss lessons for whatever reason. More than that I actually convert my notes to a PDF document at the end of each section or topic and make them available to the students to download and review outside of class. This helps the students to get away from producing 'perfect' notes and instead concentrate on following and understanding the mathematics.

## VLEs and Web Resources

The use of the internet in the classroom has exploded over the last few years, although use in the Mathematics classroom is still mostly focused on younger students, with a few exceptions. For a number of years now I have been involved in running a school Virtual Learning Environment (VLE). The focus of this work for me has been on providing resources and tools for students to use when they are outside of the classroom, much of which is based on helping them prepare for their exams. These resources take a variety of forms with the aim of making it easier for students to find the resources they need at the point at which they need it.

On my VLE at the moment I have available for my students: class notes, exam papers and mark schemes, video worked solutions to some exam questions recorded using the tablet PC or IWB with some screen recording software, handwritten (via tablet PC or IWB) worked solutions to exam papers. In addition to this students can access multiple choice questions to practice topics, and use discussion forums to ask questions of me and their peers.

The purpose of these resources is twofold; firstly to enable students access to useful resources outside of lesson time, but equally importantly it is to foster an increased degree of independence. By giving students easy access to the information they need I have found that students take more ownership over their study and are more willing to try and solve their problems themselves before coming to ask for help.

In addition to custom made web resources there are also several interesting online subscription resources such as the MEI Online Resources which attempt to cover many of the features I outlined above for a variety of A-level exam boards, or the MathsNet Resources which offer resources and activities for the A-level, Scottish Highers, IB and Pre-U syllabuses. These have the advantage of being premade and already full of interesting activities and resources, but the flip side is you lose the ability to customize the resources to your individual school.

## TI-Nspire

Another really useful tool I regularly use in my classroom in the TI-Nspire system. This system is made up of a combination of hand-held graphical calculator devices and accompanying computer software. The great strength of this platform is its ability to explore multiple representations of the same mathematical concepts. It is fundamentally a collection of linked tools, a standard calculator, a graphing package, a dynamic geometry environment, a spreadsheet, and a statistics package.

Initially I mostly used the computer software for demonstration purposes at the front of the class. For example creating a function f1 $(x)$ and then creating dependant functions such as f2 $(x)$ = f1 $(x+a)$ and using a slider to explore the effect that changing the value of $a$ has on the graph. This is a really powerful way of exploring the various transformations of functions, as it becomes simple to look at a variety of values for $a$ as well as a variety of starting functions.

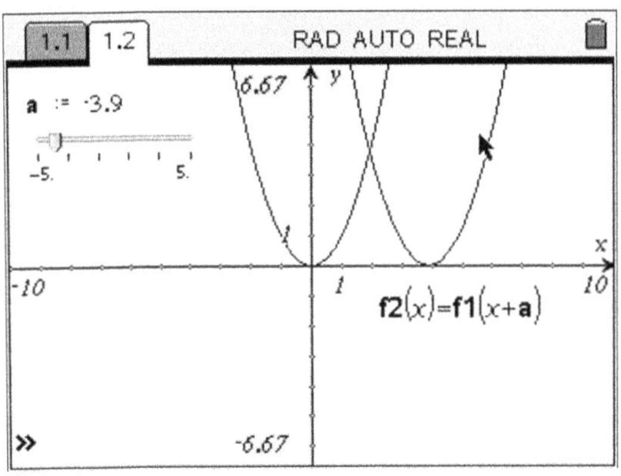

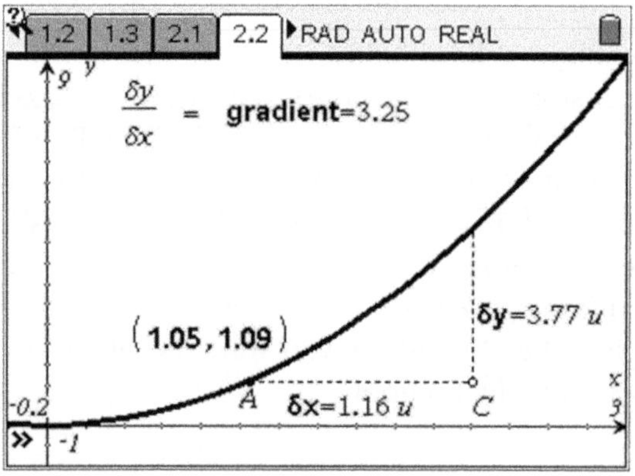


More recently I have started getting the students to regularly use the hand-held devices in class. We frequently use the TI-Nspires to get a feeling for a problem graphically before diving into the algebra. This has really helped my students get a more intuitive feel for the mathematics. Also, since the TI-Nspire has a document structure, it is possible for me to pre-create activities and copy them for students to use on the hand-held devices. Some really good examples of the kind of things that can be done with the TI-Nspire to enhance post 16 teaching can be found on the Nspiring Learning website.

## Autograph

Another technology I have used for a number of years is the excellent graphing package Autograph For post-16 teaching it is an invaluable tool for helping students to visualize much of the complex mathematics in the course. In particular I find it useful when teaching the 3D elements of the course such as Volume of Revolution. Here I am able to draw and manipulate the volume generated by rotating a graph around an axis. One of Autograph's great strengths in covering this topic is not only its ability to draw the smoothed volume but also explore what happens if you find the area under the original curve using rectangles and then consider rotating these. This helps students understand that we are really consider the sum of a series of cylinders which in turn helps them appreciate where the formula for the volume comes from. All of this can be done in just a few minutes as part of a lesson, but can completely change a student's understanding of this topic. Similarly when I introduce Conic Sections I always use Autograph to draw a double headed cone and plane and explore the shape of the intersections. The ability to freely rotate the 3D construction really helps students to visualize the ideas we are discussing.

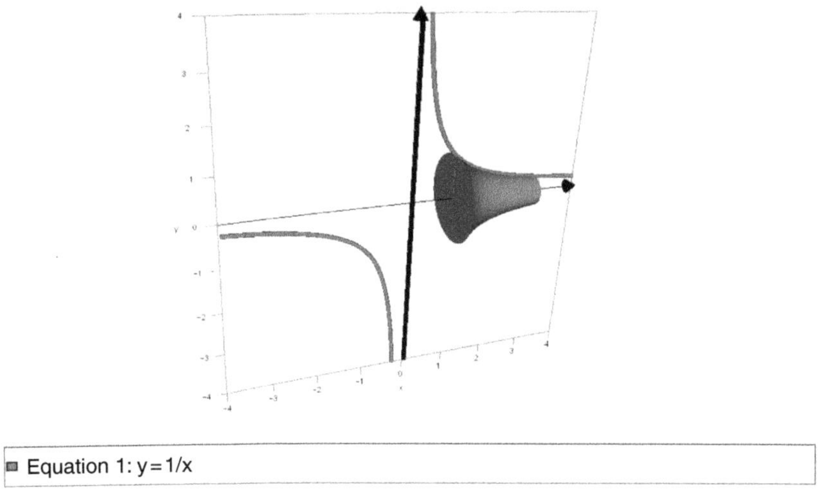


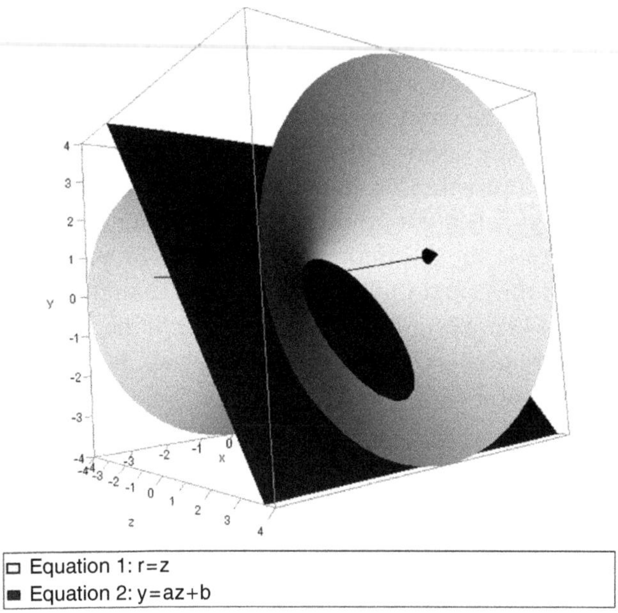


I have outlined above just a few of the ways in which I regularly use technology in my classroom and I hope you are able to see the potential impact this has on my students. For me the overriding thing that technology enables me to achieve is the ability to help students glimpse the links between the various (artificially) segregated topics in our curriculum. I want to help them see that the graphical and the algebraic are just two sides of the same coin and that algebraic trigonometry work can still be explored geometrically. It is only as students start to see these links that they realize mathematics isn't an arbitrary collection of disparate topics but rather a tapestry of interrelated ideas. For me the central reason for using technology in the classroom is to help students see the vast and beautiful thing we call mathematics.

## Conclusion

So having looked at some of the uses of technology in my classroom and identified some of the reasons why I chose to use them we must return to our starting question about whether using technology is even an appropriate thing to do in the mathematics classroom. Back in 1995, Dan Kennedy wrote the following:

> Look around you in the tree of Mathematics today, and you will see some new kids playing around in the branches. They're exploring parts of the tree that have not seen this kind of action in centuries, and they didn't even climb the trunk to get there. You know how they got there? They cheated: they used a ladder. They climbed directly into the branches using a prosthetic extension of their brains known in the Ed Biz as technology. They got up there with graphing calculators. You can argue all you want about whether they deserve to be there, and about whether or not they might fall, but that won't change the fact that they are there, straddled alongside the best trunk-climbers in the tree – and most of them are glad to be there. Now I ask you: Is that beautiful, or is that bad? (1995, 460–65)

For me the use of technology in the maths classroom enables my students to reach further and higher up into the tree of mathematics than they could by themselves, and for me that is most definitely a 'beautiful' thing!

## References

Autograph www.autograph-maths.com/ (accessed on 5 October 2009).

Kennedy, D. 1995. 'Climbing around the tree of mathematics'. *Mathematics Teacher*, September 1995, 460–65. [Online]. Available from: http://mail.baylorschool.org/~dkennedy/treeofmath (accessed on 5 October 2009).

Mathematics in Education and Industry (MEI) www.mei.org.uk/index.php?section=onlineresources (accessed on 5 October 2009).
MathsNet www.mathsnetalevel.com/ (accessed on 5 October 2009).
Microsoft Office OneNote http://office.microsoft.com/en-gb/onenote/FX100487701033.aspx (accessed on 5 October 2009).
Texas Instruments http://education.ti.com/educationportal/sites/UK/productCategory/uk_nspire.html (accessed on 5 October 2009).
TI nspiring Learning http://nspiringlearning.org.uk/ (accessed on 5 October 2009).

# Chapter 23

# Using ICT to Support Learning Mathematics in the Primary Classroom

Mel Bradford and Tina Davidson
*Elm Grove Primary School, Brighton, UK*

## Introduction

There are many Information and Communications Technology (ICT) resources available to primary teachers to support the teaching of mathematics. However there is less that supports the learning of this subject through higher order thinking skills. With that in mind we planned a series of lessons looking at the use of LOGO and the latest programmable toys in the mathematics curriculum with Year 3 and 4 children.

> In my vision, the child programs the computer and, in doing so, both acquires a sense of mastery over a piece of the most modern and powerful technology and establishes an intimate contact with some of the deepest ideas from science, from mathematics, and from the art of intellectual model building. (Papert, 1993, 2nd edition, p. 5).

Although there is some mention of programmable toys in the new Primary Framework for Mathematics, for example, Year 3 in Block D, unit 2 that was introduced in 2008, our learning objectives are taken from our school's ICT scheme from Wokingham, LA. Throughout the Primary Framework, there are numerous references to ICT resources, but there is a limited reference made to programmable toys and Logo environments, often referred to as a 'Microworld'. When referring to a 'Microworld', we are taking the definition from BECTA/DfES (2003, p 16) as 'a tiny world inside which a pupil can explore alternatives, test hypotheses, and discover facts'. On the other hand, the main ethos of the framework is that of Using and Applying mathematics and problem solving which runs throughout all the blocks and to which Logo programming lends itself as a vehicle to cover many areas of mathematics.

## Year 3 and the Use of Programmable Toys

I chose nine mixed ability Year 3 children to work in groups of three with a Pro-bot per group. Several of the children are very wary of mathematics and are reticent in taking risks with their learning. Could this be the knock on effect of ten years of the National Numeracy Strategy and a culture of testing? Although our School has an ethos of encouraging children to

learn by their mistakes and for creative thinking, I am finding an increasing number of children coming to me in Year 3, totally switched off and closed to mathematics especially when doing open ended investigations. I hoped I would be able to reverse some of these effects through using the Pro-bots, which children view as a 'toy', but through which I hoped to convey mathematical skills.

The first activity was based on getting to know how to use the Pro-bot and giving it simple instructions. Since Papert first developed the Logo programming language in the 1970s and 1980s, children's own use of computers has developed and become more intuitive as many more families have a home computer. It was pleasing to observe that all children were quickly and easily able to switch the Pro-bot on and give it simple instructions. At this stage it was very much a new toy to be played with before starting with any development of higher order thinking skills. When questioned about previous experiences in school, they had all used Bee-bots and Roamers in some context.

Giving children the opportunity to 'play' before instructing them is an important part of learning which with the pace of the primary curriculum we often neglect in favour of spoon feeding. I just happened to go into a Year 1 lesson where the teacher was instructing the children on how to turn on and program a Bee-bot. Children have been playing with toys that need to be turned on and buttons pushed to create different responses since birth. Therefore they are adequately skilled to explore the basic functions of a Bee-bot and Pro-bot, hence Bee-bots are now being used more in nursery settings.

The next task was based on estimating a distance between a fixed starting point and a fixed box. The children had to move the Pro-bot forward until it stopped just in front of the box, without knocking it over or moving it. Initially, I allowed the children to set about the task with no teacher input. It was interesting to observe Frida (a wary and weak mathematician) using her fingers spaced randomly apart at about 5cm, going along counting how many spaces this would be to the box. When questioned on her choice of non-standard measure, she was unable to give any form of answer, except a shrug of the shoulders. This adds to my belief that as teachers we have become 'spoon feeders', with the pressures of an overloaded curriculum and government targets to be met, we no longer give children as many opportunities of thinking and reflecting on their own learning and so she was unable to give any explanation. As Williams and Easingwood say 'this kind of process-orientated learning has slipped out of fashion in recent years due to the demands and indeed the misinterpretation of the National Curriculum and the National Numeracy Strategy' (2004, p. 64).

At this point, I drew the group together and asked them to show me with their fingers what Forward 1 represented. With relief I saw the majority had a spacing of a centimetre, although when asked what they were showing it ranged from 2mm to 1cm. We watched the Pro-bot draw Forward 1 and established it is 1cm. This is a great feature of the Pro-bots and different from the Roamer and Bee-bot, which can be used for estimating non-standard measures. Using the Pro-bot we can now progress to the metric system for estimating distances. The activity continued a-pace with little intervention from me, as I watched and observed some heated discussions taking place on the use of various estimates ranging from reasonable to ridiculous. We had covered estimating and measuring length earlier in the term so the children had some prior knowledge. It did seem initially there were plucking numbers from thin air. One group, whose box was about 150 centimetres away, required the command FD 150 (forward 150), but started with FD 79, then FD 100 and then totally overestimated at FD 250 before some debate about what to try next. Eventually with much refining by increasing by variables of 1 or 2 the Pro-bot came to a halt just before the box. In fact as the estimating was refined and more precise, all groups were determined to get the Pro-bot just in front of the box.

Since the children were already familiar with right angles, the following lesson built on the estimation to the box by introducing angle commands. First, pupils had to write a series of instructions to program the Pro-bot from a starting line to go around the box without touching it and return. Secondly, pupils had to refine their instructions to complete the task as quickly as possible. Having got them to write out their program before trying it, they took to the task with enthusiasm. Once again, Frida was approximating the distance with her non-standard measure! And was still unable to explain why she was doing so. It is worth mentioning that the wary mathematicians within the class, with the exception of Lee, continued to take a back seat and complained they were not given an opportunity to program the Pro-bot. They liked the idea of being in control of it, but were reluctant to be involved with the decision making process even when encouraged by other group members or myself. Others were embracing the problem solving and collaborative learning, including Lee, a very anxious and nervous child who is terrified of getting a wrong answer. Lee worked with Rob, who was taking the lead, but still listening to Lee's input and trying out his suggestions or explaining why it would not work. They easily completed the task using 90° turns and then started to refine their instructions, determined to get the Pro-bot around the box as closely as possible. Rob and Lee were able as Williams and Easingwood et al. (2004, p.109) say

> 'think Logo'. This means that they have to be able to act on their own initiatives, work as a part of a team in collaborative manner, be prepared to edit their own work and to accept that mistakes, or debugging, are an integral part of the learning process.

This is where social skills and maturity comes into play. Discussion here is about the mathematics being taught through ICT, however social interaction is an important life skill that can be enhanced through Logo environments. Where children take off with the higher order thinking skills, especially a child such as Lee, it is great to observe as a teacher but it's those children who lack the confidence that need the most encouragement and challenge us in our role. The importance of questioning comes into play and of kinaesthetic learning as a bridge to making connections. We walked through the route the Pro-bot would take, discussing if it would be a left or right turn and used a metre stick to help estimate the distance with a comparison. In retrospect, Marty and Frida had a disadvantage in that they were working with Annie, a child on the gifted and talented register for literacy but also an able mathematician, who enjoys a challenge and will easily get on with a task independently of others. Group dynamics does need careful planning and in this case did not work. Naomi and Carla worked collaboratively, making estimates and refining their answers through discussion and tried to include Laura in the process. They took longer than the others to complete the task but were eventually successful and pleased with themselves.

For the final activity I created three different tracks on the carpet for the children to program the Pro-bot to follow. All of them were used to doing right angle turns through use of the Bee-bots and previous mathematics lessons. I wanted them to explore estimating lengths and different angles of turn, hence the tracks I created had no right angles.

The estimation of lengths was far better this time, although Frida was still using her same non-standard measure. This time, it was Marty who asked her why she was doing that and explained a Pro-bot moved forward in centimetres. I later observed both her and Marty estimating a length in centimetres with far more accuracy and being more co-operative within the group. It was interesting that the third member of the group, Annie, was over-complicating the task. As I had said right angled turns were not allow, she was using multiples of 90 instead and had correctly calculated 4x90° = 360°. She was using this for an angle of turn that was slightly bigger than 90° and was confused as to why it was not working. Again I took the kinaesthetic approach and together we turned 4 right angles before repeating the same procedure with the Pro-bot and she easily realized that all she was achieving was a complete turn. Frida was then the one to suggest trying LT100 (left 100°) and from that her group were able to refine their angle of turn and continue the task.

Rob and Lee set off with determination and zeal and were soon programming lengths and angles accurately and refining them to the nearest degree. Another advantage of the Pro-bot is the visual screen where the instructions are displayed and can be followed as they are being carried out. The arrow keys then allow users to go to a line of code and edited it directly without having to retype the whole program. This is extremely useful in that it allows children to edit programs and they can immediately see the cause and effect. This makes the whole process of logical thinking and manipulation easier for them to 'hold in their heads'. The benefit for Lee, who has difficulties with fine motor skills, was the removal of the threat of having to put pen to paper to record anything. A possible disadvantage is that children are then often reluctant to record the edits to their program on paper. However, if they are refining the program and learning in the process, does recording on paper matter or is it our teacher need for evidence of achievement?

Naomi, a quiet and diligent girl, was observed counting centimetres with a finger space to find the first length. With some prompting she came up with a reasonable estimate. I put a standard 30cm ruler against the tape and from that she estimated a further 20cm. This swiftly developed into adding multiples of ten and rounding their estimates to the nearest 10 (a skill they had been learning earlier that week) before programming the Pro-bot. Carla then realized the next length was about 3 ruler lengths and added 30+30+30 mentally. Together they suggested and tried out various angles of turn.

With only one exception, Laura, all the children were working collaboratively by the last session and had achieved the success criteria and fulfilled my learning objectives. More than that, they were well motivated and eager to achieve the tasks. The looks on their faces showed how pleased they were with themselves. Through the use of Pro-bots we had covered a range of mathematical skills, including: estimation, measuring, angles, cause and effect thinking and logical thinking.

## Year 4 and the Use of Logo

Logo is the next step for children in that it is a computer program that allows children opportunities to experiment, solve problems and refine ideas on a computer screen. A small robot called a 'Logo turtle' appears on a computer screen and understands a set of written commands, such as 'RT 90' (right turn, 90°) or FD 30 (forward 30 spaces); it follows the commands to draw the corresponding lines on a screen. Children can learn to use these commands to draw simple lines or shapes. They then move very quickly to programming 'procedures' which are a set of algorithms or commands for a named pattern or shape. The simple programming language can be used effectively as a tool to further improve the way children think and solve mathematical problems as well as 'teaching the computer' to follow a set of commands.

Children at Elm Grove Primary School move into Year 4 with a wealth of experiences of programmable Bee-bots, roamers and the new Pro-bots. Their mathematical thinking is further developed by moving the children away from the physical objects to the visual movements of the screen turtle. This progress marries with the four stages of childhood development as described by Jean Piaget the eminent Swiss psychologist who Seymour

Papert worked closely with. Moving children from the Pre-operational Stage (ages 2–7) to the Concrete Stage (ages 7–11) where the child starts to conceptualize, creating logical structures to explain their physical experiences. Piaget says that 'children have real understanding only of that which they have invented themselves'. Logo allows this own experiential learning to take place. This developmental progression demonstrates the approach to Constructivism (Papert and Harel, 1991) (Resnick, 1996) based on the theories of Piaget. Constructivism 'adds the idea that people construct new knowledge with particular effectiveness when they are engaged in constructing personally-meaningful products' (Resnick, 1996).

Piaget believed that children where not empty vessels to fill with knowledge but active builders of knowledge, constantly creating and testing their new theories of the world.I was amazed when I looked closely at the Primary Framework for Mathematics 'Using and applying' objectives for Year 4, as working with Logo had the potential to cover so many objectives, from solving problems, to representing a problem as a diagram, organizing and interpreting information to find a solution, identifying patterns and relationships of properties of shapes. This was without objectives from other strands of the framework such as, understanding shapes. The ICT medium term plan for Modelling and Simulation involved learning objectives using Logo, or a Logo-type program, to write procedures to create shapes; combine procedures to produce effects including changing variables. I was therefore able to marry up the Logo modelling and simulations unit with work on angles, directions, horizontal and vertical lines and recall work on shapes and their properties from Block B unit 2 of the mathematics framework. This would also have cross-curricular links to Imaginary world's literacy unit and our main Egyptian topic.

The first session was used to reintroduce Logo and the basic commands. The children could remember FD 100 and then RT 90 for a right angled turn and quickly drew a square. They could also (much to my amazement and without any prompting) remember how to change the colour of the pen! The children worked in mixed ability pairs and were asked if they could work out how to draw other polygons. While walking around and observing I noticed that discussions on how for to turn the turtle were very well estimated, not random guesses. The children had clear ideas about 90° and 180° and made close estimates and revision of angles to make hexagons and octagons. Some pairs remembered 120° turn was needed for an equilateral triangle. Others used trial and improvement to work it out, amending and adjusting lengths of lines and angles to make their shapes possible. As the class teacher I felt my role was to prompt with questions to enable the children to find out for themselves and not give them the answers. I felt that this could lead to Logo being a more effective learning tool. Questions such as, If you know that the turtle turned a quarter turn or right angle at LT 90 is the angle you need more or less than this? If the line was FD 25 how does the next line need to be longer or shorter? Which direction will the turtle be facing when you have entered that command?

The children had not remembered the repeat function from the previous year so I reintroduced it for drawing some of the basic polygons. The children used the repeat function and shape knowledge to design and draw pictures. In the following session I wanted to introduce the 'procedure' function to draw a square and add other procedures to call them names. This was a major objective in the ICT modelling and simulations unit and involved a lot of mathematical problem solving from Block D unit 2. I modelled it on the interactive whiteboard reproducing what I had done while trying out Logo myself, firstly by writing a procedure for 'a square', then a procedure called 'line', a vertical line of squares and then a 'box'. Whilst I was modelling the process there was a lot of class discussion, the children were eager to join in and make suggestions about distances and angles for the turns and were discussing which way the turtle would point with friends. By the end of the demonstration the children saw how effectively the procedure could help them design a shape pattern, it could save time and energy of writing out several times similar commands. In the same mixed ability pairs, the children investigated writing procedures for a same shape design. Each pair could choose the shape and create their own designs. During this lesson there was a lot of discussion, pointing at screens, getting up and looking at friend's designs and talking to them about it, revising designs and changing the colour of pens. The children were extremely engaged and learning. This was an exciting lesson, the children saved their work with wonderful names reflecting the designs they made, examples were 'cool star', 'snow spin', 'colourful snowflake', 'spitfire' and 'staircase'.

From looking at examples where children printed their commands, I can see that the writing of procedures was not used by many of the groups. They worked out how to design their patterns without using this function but did have long list of commands. In future lessons I intend to continue to work with Logo, spending more time with the children learning how to writing procedures and seeing the benefit of using them. They can use this for the learning outcome of understanding that variables can be changed to allow quick changes on screen, to produce new ever increasing or decreasing shape patterns.

From using Logo and continuing to use it to complete the unit of work, I feel it has the qualities to enhance the teaching of mathematical thinking and place early concepts on angles and lengths securely in children's schemata for them to use as building blocks for future mathematical learning. I feel that the screen turtle gave an instant visual representation of their commands which would have taken a long time and probably have been beyond the average year four child's capabilities to draw. It allowed for trials to be re-worked without crossing through or rubbing out. It enabled creativity and open ended research but through a directed task. On my part as the teacher, I think the learning will only be of good quality if I continue to practise at using logo and iron out some of my lack of knowledge of its programming capabilities. I need to devote more time to careful planning, reviewing where the children have progressed to so far, in the use of writing procedures. I need to work out a few simple directed procedures so that they understand there function and capability. Then move on to changing a variable in shapes and opening up the task to allow greater flexibility of thinking and mathematical creativity. My questioning needs to be open and draw the children to greater mathematical thinking.

## Conclusion

We believe that it is our duty as teachers in an advanced technological world to educate children to use, and learn mathematical concepts through using ICT, indeed children have an entitlement to ICT in mathematics. Becta 2000 outlines five major areas for learning mathematics through ICT: learning from feedback; observing patterns; exploring data; teaching the computer and developing visual imagery. On reflection of our Logo and Pro-bot work we can see that the children have opportunities for all five entitlements. One of Logo's main strengths is that children can learn from direct and continue feedback from the visual. They are constantly making decisions about the distance for the turtle to travel, the angle for it to turn. The turtle's visual representation helps the children to 'see' the mathematical concepts and ideas and to decide what to do next. Another entitlement with ICT is 'teaching the computer'. The DFES report on school mathematics of 1985 advises that for computers to be used as powerful tools for learning mathematics, children will need to learn to program, and that

> if programming is not taught elsewhere, it should be included in mathematics lessons(DFES 1985 p. 35) Logo allows children to learn and

develop in complexity a written programming language. Howe suggests that the role of a programming language is 'to provide the learner with a 'kit of parts'. In the process of assembling the parts into a model, the learner will come to understand its structure. (Howe 1979).

We think in primary schools it is Logo's simplicity and yet diversity, from the Bee-bot to the on screen programming, that has kept it in the forefront of the mathematics curriculum as a way of introducing children to a mathematical way of thinking. It still has a standing as software that can aid learning of mathematics today.

## References

Becta. (2000). *Primary Mathematics:A Pupil's Entitlement.* http://ngfl.gov.uk (accessed on 18 March 2009).

Becta/DfES. (2003). 'ICT and pedagogy'. *ICT in Schools Research and Evaluation Series No. 18* London: DfES publications.

Department of Education and Skills (DfES). (2003). *The Big pICTure: The Impact of ICT on Attainment, Motivation and Learning* DFES/0796/2003

Howe, J. A. M. (1979). *Learning Mathematics in the Classroom through Programming: First Progress Report.* Edinburgh: Edinburgh University

Papert, S. (1993). *Mindstorms, Children, Computers, and Powerful Ideas.* New York: Basic Books.

Papert, S. and Harel, I. (eds). (1991). *Constructionism: Research Reports and Essays 1985–1990,* by the Epistemology and Learning Research Group, the Media Lab, Massachusetts Institute of Technology, Corp, Norwood, NJ: Ablex Pub.

Resnick, M. (1996). 'Distributed Constructionism'. Proceedings of the International Conference of the Learning Sciences, Northwestern University.

Williams, J and Easingwood, N. (2004). *ICT and Primary Mathematics: A Teacher's Guide.* Abingdon, Oxfordshire: RoutledgeFalmer.

Wokingham LA: Primary ICT Schemes of Work. (2009). www.school-portal.co.uk/GroupHomepage.asp?GroupId=142195 (accessed on 22 April 2009).

Chapter 24

# The Role of a Head of Mathematics Department in Ensuring ICT Provision and Use within Lessons

Dawn Denyer and Carol Knights
*Oriel High School and University of Chichester, UK*

## Introduction

Dawn Denyer is a Head of Department in a large secondary school in southern England, with a sixth form . From her point of view, the issues surrounding using Information and Communications Technology (ICT) within mathematics lessons fall into the following categories:

- Availability
- Ensuring that staff are familiar with a range of software and activities and have the confidence to use these within lessons
- Ensuring that all pupils have similar experiences within lessons – including the use of ICT

Processes for addressing these have been developed and evolved over time and also as the school has grown in size, and, although the challenges may vary in different schools, the key elements will probably be broadly similar. Ensuring that teachers are familiar with software and activities plays a major part of ensuring that pupils have similar experiences, so will be considered together.

## Availability

The availability of PCs has reduced drastically recently with the tighter controls on coursework in other subjects being implemented. This has meant that pupils need to be timetabled for sessions in computer rooms to complete their work and has inevitably led to far less 'Open Access' or bookable time being available. Thus, any time that is available to Mathematics teachers must be used as profitably as possible. In order to facilitate this and target key year groups, discussions with the person who devises the whole school timetable have proved fruitful.

The lack of access to computer rooms has forced the department to find other ways to utilize ICT and familiarize pupils with a range of software programmes. Each classroom has an Interactive Whiteboard, and one of the challenges for teachers is to ensure that they do actually use the interactive features of the boards and also involve pupils in manipulating diagrams and operating software. By encouraging pupils to have hands on access to software via an IWB, it means that when a class does have access to a computer suite, they are more familiar with the software they are using and derive more benefit from it.

Hand-held technologies provide another solution to having limited access to PCs but also have a number of advantages. The screen of a hand-held device is largely personal, whereas a PC screen can be seen by a number of other people. While this might be seen as a disadvantage, because the teacher can't *instantly* see what the pupil is actually doing on a hand-held, since most hand-helds don't have internet access, the pupil is unlikely to be doing anything other than a mathematical activity! This also allows a degree of privacy for the pupil and can be an enabler, allowing pupils to feel confident enough to experiment and make mistakes, without them being public. Hand-held technology also has the advantage of being more easily able to be an integral part of the lesson rather than the main focus of the lesson. A lesson in a computer room is often perceived as a computer lesson doing some mathematics and pupils are frequently expect to use technology for all or most of the lesson. Within the usual classroom, a mathematics lesson can take place during which some hand-held technology might be used to support learning when appropriate, which may only be for a short part of the lesson. The only negative point to be noted is that batteries often need replacing, which, while it shouldn't be a reason for not using them, is absolutely vital to the smooth running of lessons since teachers who are reluctant to use ICT are often put off by this type of problem.

## Familiarity with Software and Consistency across the Department

A Head of Department is ultimately responsible for the mathematical learning journeys of all pupils in a school. As such, Dawn feels it is important to ensure that all pupils in a year group have broadly similar experiences, including experiences with ICT.

She also has to ensure that teachers meet the new Ofsted (2009) inspections requirements; 'Resources, including new technology, make a marked contribution to the quality of learning'. Although, it could be argued that it is possible to use current software badly or to use older software effectively. The important point is to use the most appropriate software to meet the objective of the lesson, regardless of the age of the software. Indeed, there is much to be said for using familiar software with pupils in order that they can use it effectively. A number of software packages have a wide range of applications; teachers often use the same one for a variety of lessons which improves both the teacher's and the pupils' confidence with the packages and reduces the amount of familiarization time needed for both parties. A balance must be struck between the benefits of using a more current piece of software that has extra, useful features and the time spent in learning to use it. The department has an extensive range of software to draw upon and Dawn considers it is part of her role to ensure that new members of staff gradually become familiar with the nuances of the variety available.

Becta's guidance (2009) is helpful in raising teacher's awareness of the different types of usage that pupils should experience within lessons and gives examples of how these might be achieved. The guidance refers in the most part to commonly available software, which is especially helpful to the department.

To enable teachers to be discriminate in their choice of software, it is important that teachers are familiar with a range of software to enable them to be in a position to make informed decisions. As the Schemes of Work have evolved, suggestions for use of ICT to support pupils' learning have been incorporated and are regularly reviewed and updated. The Schemes also indicate when it is expected that pupils should use ICT. To support each other, the department often use regular meeting time for teacher led sessions exploring activities that could be used in forthcoming lessons, including familiarization with a range of software and discussion about how ICT might enhance activities, both within the usual classroom and in a computer suite.

Teachers within the department tend to use ICT in ways which are in keeping with their usual pedagogical approaches and within their own competencies and confidence with the software. Lessons involving ICT often require teachers to relinquish some degree of control to pupils, in order for them to explore a problem, and this can prove challenging for those teachers who like to have a stronger element of teacher control within their 'usual' lessons. Peer observation has a role to play in helping teachers develop their repertoire of pedagogical approaches, but the practicalities of organizing this often proves prohibitive. Modelling lesson structures within professional development sessions helps teachers to appreciate the power of allowing pupils to take some responsibility for making decisions within their work. The purpose of lesson modelling is twofold; empowering teachers to use software and trial activities builds confidence and also helps to ensure that these activities are used with pupils across the year group.

## Conclusion

As pupils become used to a higher level of ICT usage and graphical sophistication within society, it becomes more important to engage them through these familiar means. There are many 'visual' benefits gained through using ICT packages; mathematical diagrams can be accurately drawn for example, graphs, transformations, geometrical diagrams, and more difficult concepts can be modelled for example, velocity-time graphs and families of differential equations. This does not negate the need for pupils to be able to draw graphs or geometrical diagrams or to calculate values for continuing sequences. There is a careful balance to be struck between letting pupils manipulate scenarios created for them, say for example, to investigate circle theorems, and understanding the mathematics required to construct the scenario themselves. Using technology effectively can also allow the teacher more time to ask relevant probing questions or allows pupils more time to investigate mathematical situations without spending time on the construction of the problem.

The technology doesn't have to be the most up-to-date, it is possible to get as much out of pupils sticking matchsticks or counters on to a projected image as from using an IWB. However, software should be used when it is the most appropriate resource and teachers need to be in a position to recognize when it is. They need to see the relevance of using a particular piece of software or technology, so when opportunities arise to demonstrate this, they need to be capitalized upon.

The Head of Department has a key role to play in continuing to drive ICT use forward and ensuring that all teachers and pupils have access to, and use, appropriate resources.

## References

Becta. (2009). *Secondary Mathematics with ICT: A Pupil's Entitlement to ICT in Secondary Mathematics.* Coventry: Becta. www.teachernet.gov.uk/teachingandlearning/subjects/ict/bectadocs/sec/ (accessed on 30 June 2009).

Ofsted. (2009). *The Evaluation Schedule for Schools.* London: Ofsted.

Chapter 25

# Developing Problem Solving Skills and Cross-curricular Approaches in Mathematics Utilizing ICT

Michael Hartnell and Carol Knights
*Henry Cort Community College and University of Chichester, UK*

## Introduction

As a Head of Department in an 11–16 specialist Sports College in Southern England, Mike's teaching philosophies are to engage pupils through meaningful activity, to promote independent approaches to learning and to challenge pupils to think deeply and mathematically about substantial problems. He believes that Information and Communications Technology (ICT) has a role to play in supporting learning in a range of ways. Some of the technology he uses could be classed as mathematics software, whereas others are more generic and some are quite unusual in a classroom, perhaps stretching the definition of what is generally considered to be included as ICT use for learning in schools.

Mike uses ICT effectively in lessons to support pupils in solving problems that are real to them. These problems are drawn from familiar contexts and from the pupils' surrounding environment, which often means that the problems also have cross-curricular applicability. Problems might be quite short and fit within a single lesson, or might require a series of several lessons to solve them. Regardless of the timescale, the classroom problem solving process will often follow a similar format:

- An initial problem is posed
- Pupils gather data (and record it)
- The data are analysed
- Conjectures are made

- Conjectures are tested and rationalized relating back to the initial problem

Within different problems, Mike and his pupils use ICT to support activity at one or more of the phases. The examples that follow outline a range of problems and the software that has been used to support the different phases within this classroom problem solving progression.

One of the first cross-curricular activities that Mike used arose from a natural situation in which he wanted to give pupils a better understanding of quadratic equations, partly in response to a remark from a pupil asking what the point of the graphs were. To help pupils visualize the shape of a quadratic graph, he had the idea of taking them outside to a basketball court and videoing them trying to score a basket. He then showed the pupils how to use a free piece of software 'vidshell' together with a dynamic geometry software package to analyse their attempts, which they did in pairs on PCs.

This starting point then grew through Mike discussing the activity briefly in a staff meeting. One of the Design and Technology teachers immediately recognized links to a 'ping pong ball launcher' project and the two then collaborated, together with Science, to link the work the following year. The activity, which was also filmed by Teachers TV, enabled the pupils to make links between the subjects and transfer skills from one session to another. Software allowed pupils to not only find the equation of the trajectory but also to explore the angle of projection, to calculate maximum heights, ascertain speeds and collect data to draw accurate distance-time graphs. By capitalizing on the activity to access a range of content within different subjects, pupils gained a much deeper understanding of quadratic graphs, in a genuine context. Mike believes that carrying out activities for themselves, rather than relying on prepared video clips, helps to personalize lessons and aids pupil motivation as well as giving them a much better insight.

This activity has been used in subsequent years and the Mathematics Department timetable their part of the input to the activity to coincide in with the Design and Technology Department's scheme of work. This can be slightly problematic, but Mike argues that the benefits for the pupils, being more able to transfer skills from one subject to another and the greater insights they have, far outweigh any drawbacks. This particular activity utilized ICT at many of the phases of the classroom problem solving process: to gather and record the data, to analyse it, to assist in making conjectures and again to support in testing them.

A prime example of an activity that has promoted a much more concrete understanding of mathematics is the use of a dance mat to support learning about 2-dimensional and 3-dimensional vectors. Inspired by a local colleague who had used dance to motivate some underachieving girls, Mike brought a dance mat into the classroom. Beginning by considering mathematical notation to describe simple single moves forwards, backwards and sideways, pupils then began to build notation into their own sequences. Posing a new problem of a popular dance which required a stooping movement, pupils moved quite naturally into using three dimensional notations. The level of mathematics encountered by the pupils was far above that which they might usually achieve. Mike believes that the activity would not have been as successful if it were not for the use of a computerized interactive tool for the first phase of the problem solving process which allowed pupils to access the problem in a very natural way.

Another of Mike's strategies within lessons is to pose an interesting and non-trivial question such as 'Left-handed people are better at sport than right-handed people'. Pupils are then free to use internet search skills and any other data collection techniques appropriate to decide whether this is true or not. This is not a particularly sophisticated level of ICT use, but it is challenging for pupils to extract relevant data from the plethora available – and would be far more difficult without ICT. In an information rich society, these skills are becoming increasingly important. This activity also enables cross-curricular links, in this case with physical education, to be exploited.

Posing questions that are real and relevant to pupils is something that he encourages. With this type of question, pupils can choose to personalize the problem by considering a sport that they are more interested in. This type of activity means that the role of the teacher is altered to responding to pupil's individual questions and needs when required, and the variety of questions is more wide-ranging than if the class were all exploring exactly the same problem.

Hand-held devices such as graphical calculators are regularly used to support learning within lessons. Being portable, they are easy to integrate within lesson planning for any part of a lesson rather than having to book computer facilities for whole lessons. This allows the use of ICT to occur more naturally within the lesson rather than feeling slightly contrived, as can happen when a computer room has to be booked in advance. Building familiarity with the graphical calculators helps to develop pupils' problem solving skills; a measure of success being that pupils recognize when they will be useful in helping to solve given problems.

One unusual use of ICT that the department has pioneered, and is somewhat controversial among some staff, is the use of a set of Nintendo DS™ Originals with target groups before school. This has proved extremely successful in motivating pupils to practise basic number skills and has improved both their mathematical competence and confidence in lessons across the curriculum. Initially aimed at a Year 7 group during the summer term for one morning a week, it has grown as pupils moving on to higher year groups have insisted they wish to continue. The department currently run the sessions every school day and have pupils from Years 7 to 9 attending. The sessions are so oversubscribed that 'disappointed' latecomers have to make do with using the PCs instead of the hand-helds. Concerns raised were that the devices tend to offer drill and practice and that the 'novelty value' would soon wear off, but Mike counters this with the fact that some pupils are in their third year of attending, that drill and practice has it's place and that other games are used, particularly one that requires mathematical problem solving and logic skills. These types of games help to build a pupil's repertoire, enabling them to draw upon a wider range of skills in future.

One of the more established items of technology the department uses is the Texas Instruments Calculator Based Ranger (CBR™), an ultra-sonic motion detector which is used to explore distance-time and velocity-time graphs and provides obvious links to similar work in Science. Pupils are able to make connections between the movements they make and the 'living' graph which is simultaneously produced on the screen, giving them a better understanding of how the graph is created and what the different features of the graph relate to in real life. Mike believes this particular technology is very effective in what it does and so it still has a place within the departments' repertoire. Activities using the CBR™ can involve the use of technology in all phases of classroom problem solving, it encourages conjecturing and data can then be gathered and analysed to support or refute pupils' conjectures. It is the immediacy of the process together with the interactive nature of the activity which helps to engage and motivate pupils as well as enhance their understanding.

## Conclusion

Mike is keen for pupils to encounter a wide range of real problems and to become critically selective in the ICT they choose to use to support their problem solving efforts. This is a long-term process which requires teacher time to develop activities but also necessitates a move away from didactic teaching methods in order to facilitate pupil autonomy. Utilizing opportunities for cross-curricular activities and contexts has proven successful in improving pupils' motivation but more importantly in deepening pupils' understanding.

## References

Hard to Teach: Quadratic Graphs, Teachers.tv www.teachers.tv/video/19119

Nintendo DS, www.nintendo.co.uk/NOE/en_GB/

Texas Instruments http://education.ti.com/educationportal/sites/UK/homePage/index.html

Vidshell software www.webphysics.ccsnh.edu/vidshell/vidshell.html (accessed on 27 March 2011)

## Part Eight

# Implications for Professional Development

The subject content of the mathematics curriculum has changed relatively little in the last 25 years or so, as has, by and large, its examination and assessment system. Teachers have always been quick to embrace new technologies which make their professional lives easier, from Banda machines and photocopiers to learning platforms, e-mail and the internet. In both of these respects it would appear that serving mathematics teachers feel competent in what and how they teach – and would not identify these aspects of their work as priorities for their own continuing professional development (CPD). However it is lack of professional development, coupled with lack of time, which are usually identified as the main reasons for the very slow pace with which schools have been incorporating mathematics specific technologies into their own teaching and learning strategies. If that situation is not to perpetuate, innovative approaches to teachers' CPD and support (mutual and external) need to be adopted. This section includes contributions from a number of experts in different fields of such teacher CPD and support.

Chapter 26

# Supporting Developments within a Local Authority

Ron Taylor
*Hampshire Inspection and Advisory Service, UK*

## Introduction

This case study covers a period of almost 20 years of the Hampshire Mathematics Inspection and Advisory service. Initially Hampshire LEA included both Southampton and Portsmouth with over 120 secondary and about 800 primary schools and many other centres for nursery and special education. The introduction of unitary authorities between 1995 and 1998 reduced this but it still has some 70 secondary and approximately 450 primary schools. Providing advice and support for this number of schools over such a geographically dispersed area is a challenge. In the early years sharing and dissemination of practice was largely carried out through networks of single and cross phase groups, with meetings of 2–3 hours each half term. Over time these have been reinforced through the introduction of the Leading Mathematics Teacher (LMT) scheme, curriculum leader and professional development groups with specific teaching and learning foci. These have been of varying duration and intensity depending on the nature and focus of the group. Partnership and collaboration with higher education (HE) has been a key component of the work with schools. In the secondary sector the University of Chichester has played a significant role in the design and delivery of CPD, supported by Kings College, London, universities of Southampton and Cambridge, and the London Institute. Models of CPD which appeared to have the greatest impact, including those involving Information and Communications Technology (ICT), contained the following elements:

- Joint planning with LA inspector/adviser, secondary consultants (post National Secondary Strategy) and HE

- Interpreting and relating national priorities to school priorities
- In-school support and advice with adviser/consultant
- Time for teachers to report and reflect on the impact of new approaches and ICT tools used.

## Where Did It All Begin?

The following is an extract from the paper presented at ICTMT95 (International Conference on Technology in Mathematics Teaching 1995). It is given as an example of how Hampshire's model of CPD developed, not only in ICT but in other areas such as *Transition*, *Assessment for Learning*, and *Motivation and Relevance for 14–19 students*.

### The Hampshire Evaluation of Personal Technology in Motivating and Understanding Mathematics (The OPTIMUM)

In 1993 the Department for Education asked NCET (National Council for Educational Technology, now Becta) to manage a pilot evaluation study to assess the curriculum value of portable computers in schools in England. Proposals were invited from LEAs and teacher training institutions. A joint bid was made by Hampshire and the University of Chichester (formerly Chichester Institute of Higher Education). The main project aim was to establish:

> How ready and continuous access to portable computers could enhance pupils' problem solving skills and enhance their mathematical knowledge.

There were three sub projects in three different 11–16 schools, each with different equipment and software.

1. A portable printer and 16 laptops, with the following software:
   - MSWORKS, combined word-processing, database, spreadsheet
   - *Cabri Geometry*, dynamic geometry software
   - *Mathcad*
   - *Derive*
2. 30 personal organizers, Hewlett Packard HP95LX palm-tops with in-built graphic calculator and spreadsheet, and plug in *Derive* card.
3. Graphic calculators, Texas Instruments; 32 TI81, 14 TI85 (included link software for printing or downloading programs and data).

## Training and Support

The NCET provided all of the hardware and most of the software, however there was no provision for training, advice or support. This was funded by the LEA, HE and the schools and consisted of half a day on technical and management issues with 3 or 4 days work in schools.

In order that these new IT tools were integrated naturally into the mathematics curriculum the scheme of work was chosen as the starting point. Various aspects of content and associated mathematical tasks were analysed to establish how the particular equipment might be used to enhance pupils' mathematical learning. This often meant adapting or designing new tasks of an exploratory nature. Further technical support and training was provided when necessary. This varied according to the complexity of the machine and the 'newness' of the range of software that was available, as was the case with project school 1 when whole staff INSET was requested on the use of *Cabri* and *Mathcad.*

Staff from project 3 school also attended short courses on graphic calculators being offered to local schools by the Hampshire Maths Team.

## School Management of ICT Tools

To ensure pupils had ready and continuous access, including when necessary use at home all schools sent letters to parents outlining the schools' and pupils responsibilities.

In project 1, all pupils and staff had access to the computers. Each machine had a log book to monitor pupil usage. Pupils could also book the laptops for the completion of assignments at home. The department had a policy of encouraging pupils to independently select resources and equipment as and when appropriate. This was extended to include the newly acquired laptops.

In project 2, the personal organizers were given to a particular class of Year 10 pupils for the duration of the project and all parents were invited to a meeting to share the project aims. The organizers were carried around at all times throughout the school day and taken home.

In project 3, each pupil in one Year 10 class was given either a TI81 or TI85 for continuous use, both in school and at home. The TI85, being a much more complex machine, was given to those pupils who were considered more confident in using ICT. This left the class set of TI81s for shared use by other staff with other classes.

***Examples of Use***

*Project 1*

In the first instance much of the use was initiated by the pupils. In particular the word processing facility was immediately accessible to pupils and was used across the curriculum for writing up coursework in subjects such as science and geography. Pupils had ready access to the laptops and often took the machines home. Several, including the less able, learnt how to use the MSWORKS independently. Some utilized the spreadsheet in their geography coursework assignments.

Examples of problems chosen across the age and ability range and include:

- producing multiplication tables, use to familiarize pupils with the format and facilities of the spreadsheet;
- divisibility explorations, for example, 'when is the consecutive product of three numbers divisible by 24';
- modelling and solving resulting simultaneous equations over Integer and Real numbers;
- entering and analysing data from a variety of surveys.

Some use was also made of *Cabri*. In particular a Year 9 bottom set with a high proportion of pupils with special needs used the dynamic geometry package to investigate the ratio between opposite and adjacent sides of a right angled triangle (initially with angle of $30^0$). In the first instance, to familiarize them with the facilities of the software, they were given the task of producing a picture of their own choice such as face. Within the space of 50 minutes pupils could confidently construct circles, triangles, symmetrical points, measure angles and line segments. When constructing their right angled triangle they developed a range of strategies to create and measure the angles, perpendicular lines, and so on. Discussing with pupils the ways in which they might construct their triangle using the menu options of *Cabri* was both illuminating and challenging for the pupils.

Other classes used *Cabri* to make conjectures about a variety of angle properties in a circle such as; 'the angle at centre is twice the angle at circumference', and the special result 'the angle in semi-circle is a right angle'. In the latter stages of the project they explored other trigonometric ratios and the use of *Cabri* to develop pupils' understanding of locus and application to linkages.

*Project 2*

**Year 10 pupils.**
Pupils used a wide range of facilities available on the personal organizer such as:

- 'memo' facility for taking notes in non-mathematics lessons and for writing coursework;
- appointment diary and phonebook for working out holiday currency and imperial to metric conversions (e.g., holiday exchange rates).

The in-built spread sheet was used for:

- problem solving, modelling, solving resulting equations using trial and improvement;
- compound interest;
- simulation of dice games;
- helping with family business accounting.

The graphics calculator facility on the organizer was used for:

- normal calculations (c.f. scientific calculator but with bigger screen and ability to see several previous calculations) in maths and other subject lessons;
- graphical solution to linear and non-linear equations (only one graph at time possible, so solution of simultaneous equations required pupils to make appropriate algebraic manipulations);
- modelling problems such as 'max box' and use of solver to draw graph and find turning points.

Use of *Derive*:

- exploration of divisibility for example, for what values of m and n is $2^m+3^n$ divisible by 7;
- exploring fractional powers and surd equivalents produced by *Derive*, for example, why $99^{1/2} = 3\sqrt{11}$;
- factorizing/solving/graphing quadratic forms using *Derive* to check, manual manipulations;
- modelling for example, exploration of fairground rides such as locus of the 'Tea-Cup' ride using parametric coordinates and trigonometry.

*Project 3*

The two types of graphical calculators were used in several areas of the mathematics curriculum.

A Year 10 group taking a GCSE in statistics made extensive use of the statistical features on both the TI81 and 85. Although the TI85 was a much more complex machine and less accessible than the TI81, its facility for transferring data between machines, obtaining hard copy of data, programs and screen dumps was a bonus for pupils and staff. The hard copy was particularly useful for coursework assignments. Initially the TI85s were placed with a particular set of pupils, but once initiated on the TI81 pupils transferred from one to the other with ease. Some examples of their use included:

- Graph work: establishing equation of straight line, meaning of m and c in y=mx+c, quadratic functions arising from the product of two linear functions, relationship with their roots and graphs graphical solution of linear and non-linear equations, modelling and using graphical facilities to solve resultant equations;
- Sequences: generating number chains using a pre-loaded or given program, pupils adapting and changing the program to generate chains of their own;
- exploring geometrical designs 'Polygons and Stars' using parametric and trigonometric functions;
- Tessellations: adapting a simple program that draws a tile to create tessellations, used to help pupils understand the idea of a generalized translation;
- Statistical work: surveys use of mean and standard deviation to compare sets of data, exploring the effect of changes to data on mean and SD. Also, establishing appropriate class interval by observing changes to the histogram as the interval is altered;
- Incidental use in other subjects such as science and geography.

## Main Findings

- Frequent or continuous access to portable equipment which allowed pupils to work on problems outside normal mathematics lessons and at home led to pupils taking greater responsibility for their own learning.

They would often extend and explore new problems independently. Allowing pupils to take equipment home increased their sense of responsibility and self esteem.

- Technical support is important in the early stages to ensure the IT supports rather than hinders pupils' mathematical learning. This is particularly true with the more complex hardware/software as in projects 1 and 2.
- Some of the greatest gains in mathematical learning have been among those pupils who have had continuous access to the portables at school and at home. However, following a policy of equal access for all staff and pupils results in greater staff involvement, sharing and exchange of ideas and evolutionary integration of IT into the mathematics curriculum for all pupils, albeit at a slower rate.
- The constant feedback provided by the use of IT in mathematics allows pupils to independently explore the effect of changes they make to their model, numerical data, graphical and geometrical images. Patterns can be observed, conjectures and generalizations made and tested.
- Ensuring that pupils evaluate and make appropriate use of the IT when doing mathematics requires suitable teacher intervention and questioning.
- Even with committed staff two terms or even a year is too short a period to discern permanent benefits. The developments in the second year have been far more rapid than in the first. This is particularly true with the graphic calculators. The narrower focus and ease of use of equipment has already led to extensive use by the majority of staff and good integration into the scheme of work.
- A set of graphic calculators can be purchased for the price of a single computer. Continuous access to this power for all secondary pupils is a real possibility.
- Pupils make more informed choices on the use of IT in mathematics as they become more familiar with the range of facilities offered. This does, however, require staff to be flexible in the way in which they manage pupils' learning, in particular how they allow pupils to access resources and equipment.
- Teachers need planned support to help them explore the full potential of ICT in the teaching and learning of mathematics. It is important that teachers are given sufficient time, INSET and advice to develop the use of ICT in mathematics. This should include time for sharing ideas with colleagues in their own school and in other schools.

## What Happened Next? (Part 1)

**For Hampshire**

- The experience and knowledge gained during the pilot project was invaluable in enabling both Hampshire and the University of Chichester (UoC) to devise the joint training, advice and support programme for other Hampshire secondary schools.

**Nationally**

- Results from the pilot contributed to national CPD developments on use of ICT in mathematics.
- In 1994–1995 the DfE (now DCFS) provided monies (c£25m total with c£2m for secondary mathematics) to schools through GEST (Grants for Education Support and Training), targeted at curriculum use of ICT. Ideas, materials and approaches to teaching and learning were incorporated in materials distributed to all schools participating in this programme.
- Collaboration between suppliers, manufactures, UoC, Hampshire Mathematics Team and schools helped to inform the design and use of new ICT tools.

The first GEST ICT programme offered substantial support for hardware, software and CPD. In the UoC/Hampshire collaboration the CPD provision consisted of 5 days training, 5 days for schools to develop materials, visit other project schools and share ideas and approaches to teaching and learning, and 2–3 days in-school adviser support. There were 12 participating secondary schools with the CPD provision being spread over one academic year to enable time for trialling and reflecting upon new ideas, approaches and ICT tools. The teachers could also use their work in the project as part of a MA module.

GEST funding for ICT was also provided for 1995–1996 and a further 12 Hampshire secondary schools opted to take advantage of this provision.

## What Happened Next? (Part 2)

Outcomes and materials arising from the various projects were shared at local network meetings and the annual Heads of Department 2 day conference. There were regular collaborations with UoC and funding support from manufacturers and suppliers such as Texas Instruments through their T-cubed INSET programme. These helped to support 2–3 day training courses for about 20 schools per course over several years. Alongside these courses there were some small scale classroom based research projects. The following gives some idea of the scope of these:

- 'Data-capture and modelling in mathematics and science' funded by NCET ,
- 'Cross-curricular project' supported by the TTA (now TDA, Teacher Development Agency).
- 'Dynamic geometry project-using the TI92' supported by Texas Instruments.
- 'Linking algebraic and geometric reasoning with dynamic geometry software' funded by QCA and supported by Ken Ruthven University of Cambridge, and Adrian Oldknow, UoC, and Hampshire's leading mathematics teachers.
- 'Partners in learning cross-curricular project-uses of ICT to facilitate collaborative approaches to teaching and learning' a joint Microsoft/TDA/Hampshire Partners in Learning project.
- 'Working with 3D Geometry software in KS3 & 4' Adrian Oldknow, Ron Taylor and Hampshire leading mathematics teachers.

## Dissemination and Sharing Good Practice

The background to the structures and networks in Hampshire was described in the Introduction. It is these, along with a culture of joint and collaborative practice across schools, and between HE and the LA, that have been key in enabling the dissemination and sharing of good practice.. The involvement of the mathematics consultants, Leading Mathematics Teachers and other key teachers has increased the capacity of the LA to disseminate the teaching and learning research outcomes across all its secondary schools. In 2005 'Factors influencing the transfer of good practice' brief No. RB615, University of Sussex and international think tank Demos identified a number of key recommendations, the following is the first:

> Developing joint practice development capacity across the system (R1).
>
> LEAs should demonstrate their belief in certain kinds of collaboration. They should not only consider making joint practice development a way

of implementing a range of EDP priorities, but also make joint practice development a priority in its own right. In addition they should work with other LEAs to share their own practice. All of these activities would be of practical as well as symbolic significance.

## Conclusion

It is also worth mentioning the report on the work of de Geest et al in 2007. In it they outlined the following factors that contributed to effective CPD:

- Leadership (of the CPD);
- A practical approach;
- Stimulation, challenge and enjoyment;
- Time;
- Networking;
- Area of focus (mathematics);
- Students" learning of mathematics;
- Encouraging reflection;
- Expecting and supporting change;
- Supporting the embedding of change.

It is to be hoped that all of these are embedded within any LA and other coordinating bodies of CPD.

Chapter 27

# Supporting Teachers in Introducing New Technologies

Alison Clark-Wilson
*The University of Chichester, UK*

## Introduction

My role as a teacher, lecturer and researcher in mathematics education has enabled me to work with hundreds of (mainly) secondary mathematics teachers over the last 15 years, all of whom have wanted to develop how they make the best use of the available digital technologies in their classrooms to support their own teaching – and their students' learning. Although some of the available technologies have developed significantly in this time, for example, interactive whiteboards and classroom broadband internet access, other resources have essentially the same mathematical features such as function graphing packages. What has been most apparent in this period is that the processes through which teachers learn about and find uses for digital technologies follow some very similar patterns of behaviour, which is borne out by the substantial research that exists on this topic (Laborde 2001; Monaghan 2004; Noss et al. 1991; Pope 2001; Ruthven and Hennessy 2002).

Among the Information and Communications Technology (ICT) visionaries in mathematics education there have been some big claims concerning the way that ICT might enable us to rethink the actual mathematics we teach in schools and the experiences through which our students come to learn it (Goldenberg and Cuoco 1998; Kaput 1986; Noss 1998; Noss and Hoyles 1996; Papert 1980). The essence of their arguments concern the fact that we might spend less time in classrooms 'telling' students about new mathematical knowledge, albeit with sophisticated digital demonstrations and animations, and provide more opportunities for the students to arrive at this knowledge through exploratory tasks that use digital technologies to support students to construct mathematical meanings of for themselves.

So the first dilemma we face in thinking about how we might begin to integrate digital technologies into our classrooms concerns the following set of questions:

- Which digital technology?
- Who is going to be using it?
- When are they (or we) going to use it?
- How is it going to be used?

And behind each of these questions, what justification can we give for our decisions? (Paul Goldenberg has written a really useful article for teachers, which elaborates on these questions (Goldenberg 2000)).

The importance of your wider aim or vision for the development of your use of ICT is the most crucial factor in determining the classroom outcomes. The research evidence suggests that, where the teachers have a clear vision for their developing use of digital technology, they are in a better position to evaluate how it does (or does not) impact on their students' learning. For example, during 2007–2008, seven pairs of teachers from different English state schools began to use TI-Nspire software and hand-helds (Texas Instruments 2007) with Key Stage 3 and 4 learners. From the outset of the project they agreed their own vision for how they envisaged their students' engagement with mathematics using this technology. Cindy and Linda, who taught in a girls' school, decided that they wanted their students to use the technology to promote their mathematical discussions and began to devise lesson activities to this aim. An early lesson that they devised was Triangle angles [DCH2] which emulated a 'traditional' exploratory task whereby students draw a series of different triangles on paper, measure the interior angles and tabulate and sum each triangle's interior angles to arrive at a generalization for all triangles.

Figures 27.1 and 27.2 show the screens of the task they designed, which required the students to use the technology to measure the interior angles and then input these measurements into a spreadsheet page to observe the resulting angle sums for each of the four given triangles. The teachers were seeking to overcome the difficulties that many students have with accurately measuring angles in the paper and pencil environment.

When Cindy and Linda reflected on this lesson, they appreciated that the activity had not seemed to stimulate great mathematical discussion among

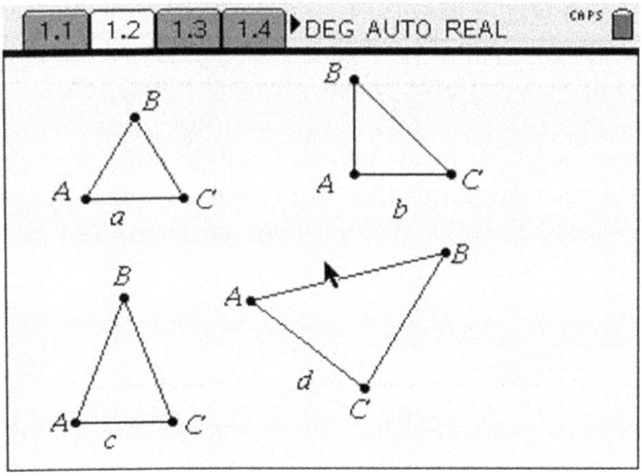


**FIGURE 27.1** [DCH2(tns-T) Page 2]

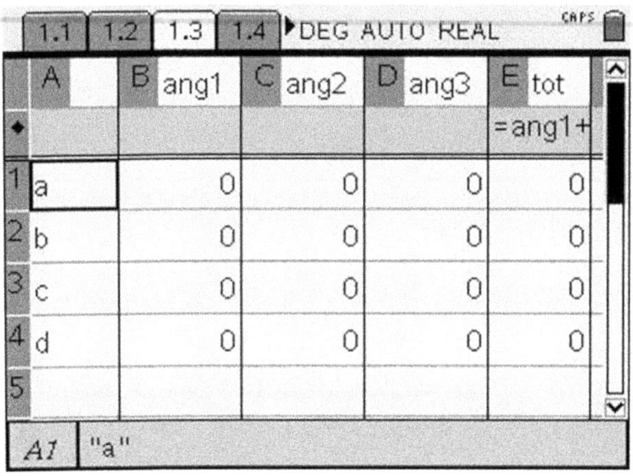


**FIGURE 27.2** [DCH2(tns-T) Page 3]

the students and we discussed this. Essentially, the students had the same four triangles and would be expected to obtain identical measurements.

They also acknowledged that, by attempting to 'translate' a familiar paper and pencil task to a technological environment, although the technology had automated the angle measuring process, they had not appreciated the power of the technology. It would have been possible for them to drawn a single triangle and, by measuring its angles, students could have explored many different cases by dragging its vertices and observing the data, captured and displayed in the spreadsheet. This is shown in Figures 27.3 and 27.4.

There are two important things to learn from this example.

1. Most teachers begin by 'translating' familiar tasks and activities to a technological environment. This is perfectly reasonable thing to do! However, if, as Linda and Cindy learned, the lesson did not seem to be

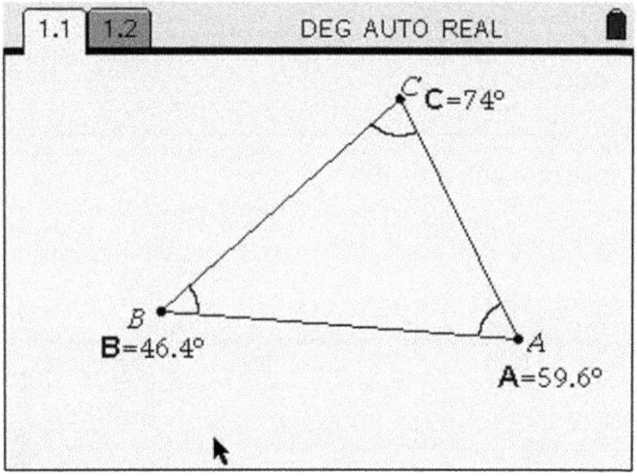


**FIGURE 27.3** A dynamic triangle

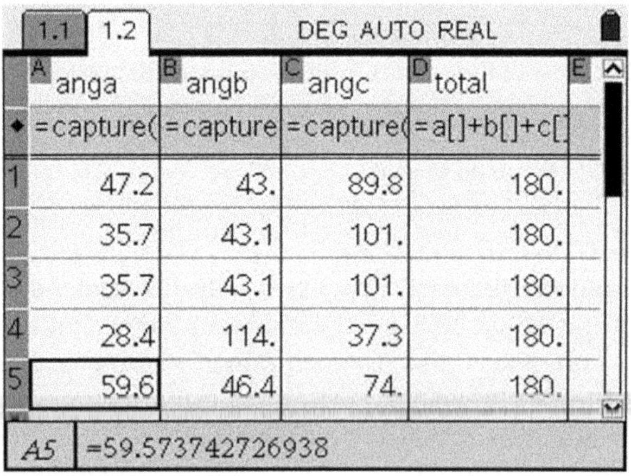

| | A anga | B angb | C angc | D total |
|---|---|---|---|---|
| ◆ | =capture( | =capture | =capture( | =a[]+b[]+c[ |
| 1 | 47.2 | 43. | 89.8 | 180. |
| 2 | 35.7 | 43.1 | 101. | 180. |
| 3 | 35.7 | 43.1 | 101. | 180. |
| 4 | 28.4 | 114. | 37.3 | 180. |
| 5 | 59.6 | 46.4 | 74. | 180. |



**FIGURE 27.4** The captured angle data

worth the effort, many teacher then give up on the technology and revert to the less risky approach which is the paper and pencil one or choose to demonstrate the activity to the class themselves. By rethinking the task completely, a more engaging lesson was developed which enabled the students to have a hands on experience.

2. Cindy and Linda needed to know that it was possible to drag the triangle dynamically and measure its angles automatically in the spreadsheet. This required them to develop their own personal skills with the technology in a supported environment – something that would normally happen in some form of professional development opportunity.

## Conclusion

Learning to use technology for teaching is not the same as learning to use technology for ourselves. There is an additional layer of consideration which involves us thinking about underlying pedagogy we want to adopt and the way in which we select or design a task for our students. Knowing what is possible with the technology changes the decisions that we make.

## References

Goldenberg, P. (2000). 'Thinking (and talking) about technology in math classrooms'. Available online at: www2.edc.org/mcc/pdf/iss_tech.pdf (accessed on 14 December 2009).

Goldenberg, P. and Cuoco, A. (1998). 'What is dynamic geometry?'. In R. Lehrer and D. Chazan (eds) *Designing Learning Environments for Developing Understanding of Geometry and Space*, pp. 351–67. Hillsdale, NJ: Lawrence Erlbaum.

Kaput, J. (1986). 'Information technology and mathematics'. *Journal of Mathematical Behavior*, 5 (2) 187–207.

Laborde, C. (2001). 'Integration of technology in the design of geometry tasks with cabri-geometry'. *International Journal of Computers for Mathematical Learning*, 6 (3), 283–318.

Monaghan, J. (2004). 'Teachers' activities in technology-based mathematics lessons'. *International Journal of Computers for Mathematical Learning*, 9, 327–57.

Noss, R. (1998). 'New numeracies for a technological culture'. *For the Learning of Mathematics*, 18 (2), 11.

Noss, R. and Hoyles, C. (1996). *Windows on Mathematical Meanings: Learning Cultures and Computers*. Dordrecht, The Netherlands: Kluwer Academic Publishers.

Noss, R., Sutherland, R. and Hoyles, C. (1991). 'Final Report of the Microworlds Project Vol II: Teacher attitudes and interactions'. London: Institute of Education.

Papert, S. (1980). *Mindstorms: Children, Computers, and Powerful Ideas.* New York: Basic Books.

Pope, S. (2001). *Handheld Technology Project.* Coventry: British Educational Communication and Technology Agency.

Ruthven, K. and Hennessy, S. (2002). 'A practitioner model for the use of computer-based tools and resources to support mathematics teaching and learning'. *Educational Studies in Mathematics,* 49, 47–88.

Texas Instruments. (2007). TI-NspireTM, Dallas, TX: Texas Instruments Inc..

Chapter 28

# Implications for Professional Development: Supporting Individuals

Pip Huyton
*NCETM Consultant*

## Introduction

Mathematics teachers today are expected to keep up to date with current educational initiatives, pedagogical thinking and to have sound mathematical knowledge. The most recent revision of the National Curriculum, to one based upon key processes and concepts, is one example of the challenges for teachers to rethink the way they and their learners interact in the classroom. Traditionally, professional development for teachers has been provided through courses; many of which occurred during the school day, were held in venues determined by the providers, and were limited in their potential for continuous development. The introduction of digital technologies has provided alternative methods of communication which allow opportunities for teachers to engage in continual professional development 24/7 and for 365 days of the year – providing the personalized flexibility in how, when and where the professional development is accessed, via their PC with internet connection. All of the new technologies mentioned here are freely available via the internet and are therefore accessible for all teachers at all times.

Digital technologies which have, in my experience in working with individuals and networks of teachers, been significantly useful in supporting the contemporary professional development of teachers may be classified as the 3Vs. They are

- Video Capture
- Virtual Real Time meetings
- Virtual Communities.

## Video Capture

One example of video capture software is *Jing*, a TechSmith free download: www.jingproject.com.

For many individuals a multi sensory approach can enhance their learning experience. For the teachers' professional development to have an impact in the classroom the learning experience needs to be positive and supportive. A video capture includes pictures, sounds and words which together create a detailed explanation of an activity; providing an experience which the teacher can then personally put into practice.

The software allows the user to record events that occur on the desktop of their PC and then save the recording either as a file onto a drive of the PC or to a TechSmith server which then generates a web link. There then exists an opportunity to share the file with other teachers in the department, the locality or the wider mathematics community. The video capture is most effective when it is brief, focussed and a snapshot of an event. Experience has shown it has great potential in supporting the development of Information and Communications Technology (ICT) in the teaching and learning of Mathematics.

A '*how to*' video capture has been a particularly useful application of the technology. For example, a recording of how to use specific features of mathematical software for teachers could be a development of their own skills in using the software, and also can be used as an aide-memoire for students working with the software in the classroom. Being able to replay (and pause) the video as often as required-with instant access supports the individual at all times. Inserting the link to the video capture in any relevant electronic documents establishes an immediate connection if required; teachers found this helpful when preparing materials for students of differing abilities.

A 'this is how we did it in the classroom' video capture is another positive use of the technology. Teachers are able to easily narrate recordings that demonstrate how an activity developed in their classroom as an example to share with others and jointly analyse. The recording can demonstrate the questions asked, whether oral or written, the sequence of the events, the responses from learners and the way in which the software aids the learning. The power of this recording for professional development is in its analysis and consideration of what else might have been done. For example, what other questions might the teacher have asked? Would the learning experience be improved by re-ordering the events? How could we engage the learners more? The advantage of this type of video capture is that it is a brief resume of a lesson which teachers can find time to watch.

## Virtual Meeting Rooms

One example of a free virtual room is provided by *Elluminate*: www.elluminate.com.

In the virtual meeting room teachers can log in at an agreed time to 'meet' their colleagues or learners without having to move away from their internet connected PC. In built features such as webcam and audio allow users to see, hear and speak to each other on-line, while screen sharing enables users to view and, with permission of the user, take control of other users' desktops. The combination of features therefore provides a platform for genuine distance learning and communication. Teachers have been able to bring colleagues and learners 'virtually' into their room. The real time experience allows for questions and answers to be instantaneous within the group of users.

The 'never too far' virtual meetings are particularly beneficial for teachers from different schools wishing to work collaboratively on an activity. Quality resources are often a by-product of professional development, being created as a result of teachers' ideas being expanded, trialled in the classroom, reflected upon and then appropriately modified – this requires repeated communication between teachers developing such resources. Being able to 'virtually' meet facilitates the collaboration of the teachers on a more frequent or regular basis, particularly when the location of the teachers is dispersed. In the virtual room, all teachers simultaneously view the same item, allowing for questions like 'can you just explain' and 'can you show me how you did' to be immediately answered, before the thought is lost. Documents can be annotated, amended and then saved for all users to access.

The 'have you thought of' virtual meetings enable teachers to directly link to a colleague with a different skill set or perspective, which may result in the exploration of new ideas and methods or improved subject knowledge. This was the case when small groups of students, supervised by their member of staff, 'virtually' met as part of a G&T student network. An outcome of the meetings was that peer to peer learning took place which included one group demonstrating how, in that particular case, Dynamic Geometry software allowed the students' mathematics to be developed further than the pencil and paper method used by another group. As a result of this interaction one of the teachers identified a specific PD need, namely how to use the software that was already installed on the school network. In addition the teacher had established an initial point of reference for professional advice.

## Virtual Communities

The web portal of The National Centre for Excellence in the Teaching of Mathematics (NCETM) www.ncetm.org.uk is one free provider of a platform for collaboration through its communities.

Online communities, or forums, provide a virtual space in which teachers can observe, learn and share. Teachers can engage with each other, and the forum facilitates networking. Teachers can ask and respond to questions; contribute to the making and sharing of resources; challenge and be challenged in their mathematical knowledge and pedagogy. This can occur with an invited group of mathematicians in a closed/hidden community or with the wider mathematics society in an open community.

## Conclusion

A '*what makes it work*' community encourages teachers to reflect on their practice, analysing all aspects of their role in the classroom for example, what they did, what they said, who they worked with and what equipment was used. The outcome of engaging with a community for teachers is a greater appreciation of the variety of practice beyond their own school. This in most cases motivated teachers to step out of their comfort zone and to try alternative methods, which increased their personal confidence. Examples of this are communities that worked with the materials developed, under the leadership of Malcolm Swann, entitled 'Improving Learning in Mathematics'. The individual teacher can access a community and within that community identify or create a topic of discussion linked to a key personal interest. Comments posted within the topic may assist the teacher in obtaining helpful advice or possible solutions to issues such as 'how can I use the IWB more interactively in the classroom?' or 'how can I adapt an activity to work with only one computer in the classroom?'

Through a combination of the 3Vs teacher can be facilitators and learners within any collaborative community that provides access to regular CPD.

Finally, while there are significant advantages to be gained by using the new technologies as part of the teachers' professional development the element of personal interaction of the traditional methods for CPD is still valued.

Chapter 29

# What Are the Significant Factors Which Support the Integration of ICT in the Mathematics Classroom?

David Wright and Pam Woolner
*Research Centre for Learning and Teaching (CfLAT), Newcastle University, UK*

## Introduction

The introduction of interactive whiteboards into classrooms has seen an increasing use of Information and Communications Technology (ICT) by teachers. However, this may mean that students are getting fewer opportunities to use ICT individually or in small groups (Smith et al., 2006). Where this does happen, it frequently involves booking a special 'computer lab', reducing the frequency and 'ad hoc' use of ICT as part of the mathematics classroom.

> Several years ago, inspection evidence showed that most pupils had some opportunities to use ICT as a tool to solve or explore mathematical problems. This is no longer the case; mathematics makes a relatively limited contribution to developing pupils' ICT skills. Moreover, despite technological advances, the potential of ICT to enhance the learning of mathematics is too rarely realised. (Ofsted, 2008) p. 17.
>
> In the secondary schools, the two main problems were the lack of ICT resources and weaknesses in identifying suitable activities at key points in schemes of work. (Ofsted, 2008) p. 18.

We investigated the impact on teachers and students of the introduction of small software applications on hand-held technology (graphical calculators or GCs)in the mathematics classroom. The study (Wright and Woolner, 2009), supported by a grant from NCETM, was a two year enquiry which began in June 2007 with the delivery of the GCs and was completed in March 2009.

## Background

We believed that graphical calculators could be useful ICT tools which can be put into the hands of every student in a mathematics lesson without having to take the class off to the ICT suite. This seemed appealing because the calculators are able to run versions of the small software, sharply focused on specific topics, which is established as a useful resource to support learning in mathematics. Specific examples include the SMILE software (www.bebc.co.uk) and ATM programs (www.atm.org.uk).

As this demonstrates, graphical calculators can be used for rather more than graph-plotting (Hennessy, 1999; Wright, 2004) and recent technological developments also make it possible to network calculators together, allowing pupils and teachers to share an interactive space.

The teachers were supplied with a GC emulator (*TI Smartview©*), which allowed them to model the GC on their PC and project the image on to a whole class display, facilitating explanations and the modelling of procedures. A later addition to the project is a networking system, the TI Navigator©, which allows the GCs to be linked to each other and to the teacher's PC so that data and images can be shared and 'screen shots' from the GCs projected on to the whole class display.

The initial approach to the use of the technology via small software was chosen as the most accessible and motivating route into its use. This approach was chosen since we hypothesized that the adoption of new technology by teachers and learners is most successful where it does not involve a big commitment initially in learning about its functionality, where the application fits in well to teachers' existing practice and where there is an immediate gain in 'value added' to the learning of the students. However, we were aware that this approach may lack challenge and fail to have an impact on classroom practice. Thus some episodes of outside training were also offered.

The two mathematics teachers centrally involved in the project were interested in the potential of the GCs, but had no previous experience of using such technology in their teaching. They were given GCs and some initial training (mainly focused on how to load and run small software on the GC) in the summer term 2007, then used them in class from September 2007. The Navigator networking device was delivered to them during this first term and they received training on it in February 2008 and began to use it in their lessons from that date.

Throughout the project, we worked with the teachers and students to investigate their experiences of using the hand-held technology for mathematics teaching and learning. Data was gathered through informal interviews and meetings, classroom observation, teacher diaries and pupil questionnaires.

## Learner's Perspective

There was an initial, consistent perception among the learners that the use of the GCs in their lessons was beneficial. This impression was supported by observation of lessons, including informal comments made by the students, and corroborated the opinion of the teachers that the GCs were generally well received by their classes.

It was anticipated that the GCs would have a wider impact on learning, perhaps stimulating problem solving and higher level talk around mathematics. During lessons observed early on in the project, much of the talk between teacher and learners and between learners seemed to concern the practicalities of using the GCs (e.g., 'Where's the timer?') or specific, fairly low level requests for help (e.g., 'What's 0.98 x 9..?'). During later classroom observations, a higher proportion of comments related to mathematics were noted. For example, in a lesson using the graphical functions to explore equations of lines, learners were heard discussing each others' graphs, asking nearby students 'How did you get that?' and offering ideas to the teacher:

> 'If you divided by 2 would it be, like, the other way round?'

These observations support the frequently reported impact of the effectiveness of ICT in promoting conjecturing and experimentation, yet here they are taking place in the context of the normal mathematics classroom and lesson, not in the ICT suite.

As mentioned above, there are concerns that although ICT is present in the contemporary mathematics classroom it is often limited to the IWB, and so mainly used by the teacher. In this respect the provision of hand-held technology for learners makes a clear difference. The pupil questionnaire responses demonstrated that, from a fairly early stage, the GCs were being used for a variety of mathematical topics. Even when the use of the GCs was not particularly mathematically sophisticated, the learners seemed motivated, enjoying the challenge of, for example, problem-setting programs:

> 'timing is fun against friends'

Later lessons involved the use of the Navigator networking device. This use of the Navigator produced some instances of 'interactive whole class teaching', as well as learning, which seemed generally to engage all the learners. The questionnaire responses show that the students found the mathematics activities which used the Navigator more memorable than those that did not.

## Teacher's Perspective

The teachers' initial reactions were favourable and the teachers found the training on how to access the software particularly useful.

> Very helpful session, I feel less intimidated at the thought of using the calculators in the classroom with the students. The programs we saw were excellent and I am looking forward to using them with the students.
>
> seems at the moment to be a wonderful resource!

The potential for students to progress at their own pace was noted;

> it was excellent for the higher ability pupils especially as they could progress at their own speed

It also seemed to motivate the students to work on what could be quite 'dry' material (arithmetical operations):

> pupils probably did more examples than they usually would have.

The unfamiliar resource also caused some problems:

> I am not able to get round the room quickly enough to support each pupil as soon as they need help. Extra staffing would be nice.
>
> I felt the lesson was a bit stressful as some pupils could not progress due to technical problems.

This suggests that extra support should be made available when GCs are introduced so that teething problems can be quickly dealt with. However, these issues were outweighed by the initial positive impact in motivation and enjoyment in learning afforded by the introduction of the GCs.

> I feel [the] motivation of [the] class and overall pupil enjoyment of maths has increased dramatically using the calculators. Especially effective for engaging boys who sometimes do little work in the lesson!

Teachers quickly noted that there was a need for another style of pedagogy in using this facility.

> I found the program is good for testing understanding, but I found it hard to assess pupils understanding during the lesson due to pupils being asked different questions.

Note though, that in contrast to a similar exercise using written work, the students are no longer able to give each other 'the answer' without engaging in some level of problem solving. Further work with students to reflect on how engage in peer support in these situations could increase this level of interaction.

Later diary entries, particularly in the class where the students have ownership of the GC, suggest that the initial novelty has 'worn off' so that more attention needs to be paid to the appropriateness of the activity.

> The lesson was ok, but I'm now feeling that familiarity is breeding not contempt, but certainly the novelty is wearing off and some pupils are not working as hard as they could be. [However, it should be noted that a classroom observer comments that the proportion of on-task behaviour generally remains 'pretty high' even when there are problems.]

Additional benefits of access to this resource that were noted included the flexibility afforded so that students who had been absent could be introduced to the resource and catch up at their own pace.

There was some additional difficulty in developing the use of the GC as a 'tool' which supported the students' thinking, because both the teachers and students were unfamiliar with this aspect of the technology. However, many students remarked on the graph plotting facility as something they recalled from using the GC. The technology also had its own limits, for example, it was not possible to plot the graphs of 'implicit' functions without students transforming them into the 'y = ' format which caused extra problems when students were unable to perform the transformation accurately.

As the project progressed the teachers began to be curious about the range of functionality of the GCs and the potential of the networking facility.

The training offered in February addressed both of these issues and introduced the teachers to some ready-made classroom activities which could be used with minimal development. This input was very well received:

> We were 'fired up' again, and were impressed with how the calc's can be used to explore accuracy, fractions, standard form and algebraic identities.

## Conclusion

There are a range of issues for further professional development here. Evidence from the teachers' diaries noted that, despite their growing familiarity with the GCs, the learners' response varied widely according to the choice of activity and resource used. Some activities were extremely popular and students returned to their use voluntarily either in the classroom or at home. Hence there is a need for both for teachers to become more informed about the potential pedagogical value of particular activities and also to be able to judge what that potential might be in the context of the GC and how and when to mediate and intervene in the students' activities.

The teachers are developing their experience in how and when the use of this technology might be appropriate. Their diary entries demonstrate that, despite the immediate, generally positive, impact of the technology, professional judgement and experience was being developed to support decisions about when, how and whether to use this resource.

The combination of hand-held personal technology and a networking system combines the advantages of the computer lab and projection technology to create a shared interactive space (digital workspace) in which to do mathematics in the ordinary mathematics classroom. (Davis, 2003, Hivon et al., 2008) In this context ICT can be used when the teacher and/or students choose to do so. In addition, students and teachers can share their work or even all participate in the same activity where everyone can see the whole class's contribution. The 'connected classroom' also crucially supports the assessment and feedback process essential to supporting learners' development.

### What Has Been Learned?

1. Early adoption of technical innovations by 'mainstream' teachers depends on:

- a relatively undemanding commitment initially in learning about its functionality;
- the application fitting in well to teachers' existing practice;
- a perceived immediate gain in 'value added' to the learning of the students;
- readily available technical support to sort out any 'hitches';
- including an 'outside' influence to provide an initial 'boost' and to sustain the promotion innovative activities.

2. Planning: both 'large scale' schemes of work and 'small scale' lesson planning is crucial for the sustained use of the technology. This has implications for the allocation of time for professional development.
3. If teachers have a sense of ownership of the innovation and students have personal ownership of the technology, the innovation is more likely to be sustained.

## References

Davis, S. (2003). 'Observations in classrooms using a network of handheld devices'. *J. Comp. Assisted Learning*, 19 (3) 298–307.

Hennessy, S. (1999). 'The potential of portable technologies for supporting graphing investigations'. *British Journal of Educational Technology*, 30 (1) 57–60.

Hivon, L., Pean, M, Trouche, L. (2008). 'From a network of calculators to collaborative construction in the class'. Lyon, France, University of Lyon. *Repères-IREM (2008,n°72)* downloaded at http://ti-researchlibrary.com/Lists/TI%20Education%20Technology%20%20Research%20Library/Attachments/76/CROME-Reperes_english.pdf (accessed on 6 February 2011).

Ofsted. (2008). *Mathematics: Understanding the Score* HMI 070063.

Smith, F., Hardman, F. and Higgins, S. (2006). 'The impact of interactive whiteboards on teacher-pupil interaction in the national literacy and numeracy strategies'. *British Educational Research Journal*, 32 (3) 443–57.

Wright, D. (2004). 'Graphical calculators: tools for thought'. In Johnston-Wilder, S. and Pimm, D. (ed.) '*ICT for Secondary Mathematics*' Milton Keynes: Open University Press.

Wright, D. and Woolner, P. (2009). *Small Software for Mathematics on Hand Held Technology*. Downloaded at www.ncetm.org.uk/enquiry/5166 (accessed on 10 August 2009).

# Glossary

| | |
|---|---|
| AT | Attainment target see MA1,2,3,4 |
| ATM | The Association of Teachers of Mathematics |
| avi | Audio video interleave. A multimedia container format introduced by Microsoft for video files |
| Becta | British Educational Computing and Technology Agency formerly NCET |
| BIS | Department for Business Innovation and Skills |
| bmp | File format used to store bitmap digital images |
| CAS | Computer Algebra system |
| CBL | Calculator based laboratory |
| CBR | Calculator based ranger |
| CD-ROM | Compact disc containing Read only files |
| Curriculum dimensions | 7 Cross curricular themes<br>Identity and cultural diversity<br>Healthy lifestyles<br>Community participation<br>Enterprise<br>Global dimension and sustainable development<br>Technology and the media<br>Creativity and critical thinking |
| D&T | Design and technology |
| DCSF | Department for Children , Schools and Families, formerly the DES |
| DES | See DCSF |
| DGS | Dynamic Geometry software |
| DIUS | Department for Innovation, Universities and Skills now part of BIS |
| GC | Graphing Calculator |
| gif | Graphics Interchange format a bitmap format used for simple usually graphic images |
| GSP | Geometer's sketchpad |
| ICT | Information and Communication technology |

| | |
|---|---|
| IT | Information technology |
| IWB | Interactive Whiteboard |
| jpeg | Commonly used method of compression for digital photographic images. The Joint Photographic Experts Group standard, |
| KS1 | Key stage 1: 5-7 years |
| KS2 | Key stage 2: 7-11 years |
| KS3 | Key stage 3: 11-14 years |
| KS4 | Key stage 4: 14-16 years |
| KS5 | Key stage 4: 16-19 years |
| LAN | Local Area Network |
| LCD | Liquid crystal display |
| MA | The Mathematical Association |
| MA1 | Attainment target 1 Using and Applying mathematics |
| MA2 | AT 2 Number and Algebra |
| MA3 | AT 3 Geometry and Measures |
| MA4 | AT 4: Statistics |
| MB | Mega Byte. A measure of file size equal to 1 million Bytes or characters. |
| Mb | Megabit 1 million bits. One byte equals 8 bits (binary digits). Download speeds are given in Mbps or megabits per second. |
| NANAMIC | National Association for Numeracy and Mathematics in Colleges |
| NCET | See Becta |
| NCETM | The National Centre for Excellence in the Teaching of Mathematics |
| NCTM | The National Council of Teachers of Mathematics (US) |
| NLVM | The National Library of Virtual Manipulatives at Utah State University |
| OCR | Optical character recognition |
| OFSTED | The Office for Standards in education, children's services and skills |
| OHP | Overhead projector |
| OS | Operating system e.g. Linux, Windows XP, Vista... |
| PC | Personal computer |
| PCMCIA | Personal computer memory card international association. These cards are used for computer memory storage expansion |

| | |
|---|---|
| pdf | Portable document format – a file format developed by Adobe Systems |
| PLTS | Personal learning and thinking skills<br>Independent enquirers<br>Creative thinkers<br>Independent learners<br>Team workers<br>Self-managers<br>Effective participants |
| QCA | See QCDA |
| QCDA | The Qualifications and Curriculum Development Agency, formerly QCA |
| RAM | Random Access memory |
| ROM | Read only memory |
| SSAT | Specialist Schools and Academies Trust |
| STEM | Science, Technology, Engineering and Mathematics |
| swf | Small web format. A file format used particularly for animated vector graphics. Associated with Shockwave Flash and Macromedia and Adobe, |
| TDA | The Training and Development Agency for schools, formerly the Teacher Training Agency (TTA). |
| tiff | Tagged image file format. A file format for storing images particularly for image manipulation operations and OCR. |
| TIN | TiNspire handheld device and software |
| TTA | See TDA |
| UMPC | Ultra mobile PC |
| URL | Uniform Resource locator – web address |
| VDU | Visual display unit |
| VGA | Video Graphics Array<br>VGA connectors are used to connect PCs to data projectors external monitors etc |
| VLE | Virtual learning environment |

# Index

www.ingramcontent.com/pod-product-compliance
Ingram Content Group UK Ltd.
Pitfield, Milton Keynes, MK11 3LW, UK
UKHW020135290926
15471UKWH00003B/507

9 780567 25028